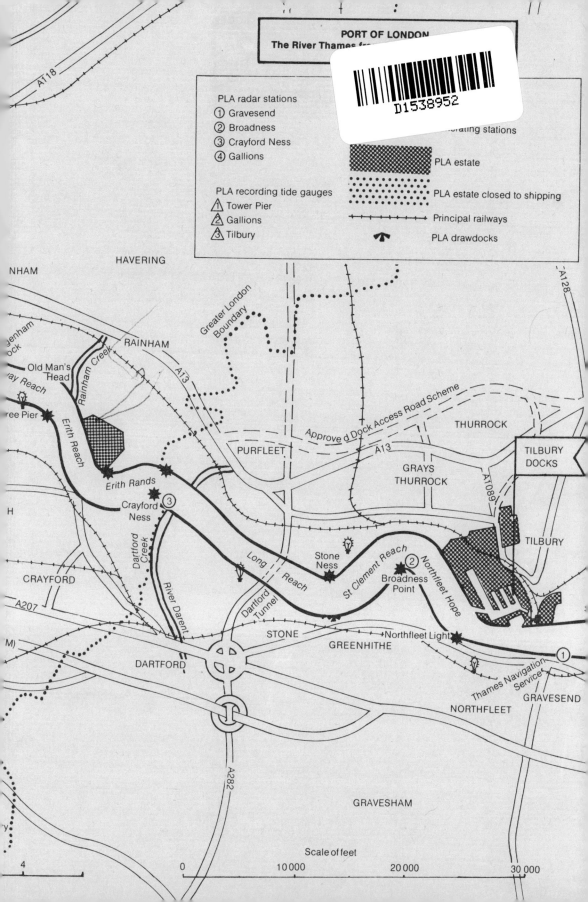

PORT OF LONDON
The River Thames f...

PLA radar stations
① Gravesend
② Broadness
③ Crayford Ness
④ Gallions

PLA recording tide gauges
⚠1 Tower Pier
⚠2 Gallions
⚠3 Tilbury

...rating stations

▓ PLA estate

▒ PLA estate closed to shipping

┼┼┼ Principal railways

⚓ PLA drawdocks

HAVERING

NHAM

A18

Greater London Boundary

RAINHAM

Old Man's Head

ray Reach

ree Pier

Erith Reach

Rainham Creek

A13

Erith Rands

Crayford Ness ③

Dartford Creek

CRAYFORD

A207

River Darent

PURFLEET

Approve d Dock Access Road Scheme

A13

THURROCK

GRAYS THURROCK

TILBURY DOCKS

A1089

TILBURY

A128

Long Reach

Dartford Tunnel

Stone Ness

St Clement Reach

Broadness Point ②

Northfleet Hope

H

M)

DARTFORD

A282

STONE

GREENHITHE

Northfleet Light

NORTHFLEET

Thames Navigation Service

GRAVESEND ①

GRAVESHAM

Scale of feet

4

0   10 000   20 000   30 000

London Docks 1800–1980

a civil engineering history

Building St Katharine Dock

London Docks 1800–1980

a civil engineering history

IVAN S. GREEVES, FICE

WITHDRAWN

London : Thomas Telford Limited

First published 1980 by Thomas Telford Limited, PO Box 101, 26–34 Old Street, London, EC1P 1JH
Distributed by Thomas Telford Limited, 1–7 Great George Street, London, SW1P 3AA

CONVERSION TABLE
As this is an historical book, most dimensions are quoted in Imperial units. After 1973, the units are metric, reflecting the general changeover around that time.

1 in.	25.4 mm
1 ft	304.8 mm
1 yd	0.9144 m
1 mile	1.609 km
1 sq. in.	645.16 mm^2
1 sq. ft	0.0929 m^2
1 cu. ft	0.028 m^3
1 cu. yd	0.764 m^3
1 acre	0.405 ha
1 cwt	50.8 kg
1 ton	1.016 tonne
1 lb	0.454 kg
1 lbf/sq. in.	6.895 kN/m^2
1 cwtf/sq. ft	5.362 kN/m^2
1 gallon	4.54 litre
1 hp	745.7 W

ISBN: 0 7277 0114 2

Photoset, printed and bound in Great Britain by REDWOOD BURN LIMITED Trowbridge & Esher

Preface

In this book I have tried to record the major engineering developments carried out in the Port of London over a period of 180 years. Since 1943 I have spent over 30 years as an engineer working in the port and thus have had first hand experience of many of the difficulties confronting engineers in the design, but more especially in the construction, of dock installations.

The ever-changing demands made by shipping companies can lead to expensive new works being undertaken with no guarantee that when they have been provided, they will be used. If the engineer thinks big he is accused of extravagance: if he does not plan far enough ahead the ships will go elsewhere; he can seldom be right.

The decision to provide London with enclosed docks proved to have both its advantages and limitations, particularly with regards to the Upper Docks. They provided a safe unloading point for a flood of imports and exports resulting from free trade and the industrial revolution. With the loss of their trade monopolies, however, a bitter rivalry and cut-throat competition led to ruthless and uneconomic price cutting, including reductions in wages, leaving a bitterness still influencing labour relations 150 years later. (The subject of labour relations is outside the scope of my book.)

The chaos resulting from the private companies' inability to maintain safety standards or provide money for new construction led to the formation of the Port of London Authority (PLA), which was to weld the warring factions into an organization capable of providing a viable port. The new organization was fortunate in having at its head two forceful personalities in its Chairman, Lord Devonport, and Chief Engineer, Sir Frederick Palmer. With considerable foresight they planned together the construction of what was to become the greatest port in the world. Their plans were carried out until stopped by World War II, which brought its own problems and achievements, including the use of the port for the construction of many of the Mulberry Harbour Units.

Post-war reconstruction of the bombed Upper Docks led to many difficulties, though it was at this time that the port reached its busiest: between 1959 and 1962 every berth was occupied and there were queues of ships off Southend waiting to unload cargo.

At the same time the enclosed docks close to the City of London which had given the port its early start were now limiting it. The 'container revolution' and new ways of handling timber led to the building of larger and larger ships which were unable to enter the Upper Docks. The centre of activity moved to Tilbury, where development started in the 1880s but had been overshadowed by the facilities upstream until the 1960s. The major developments carried out at Tilbury during the last 20 years have left the Upper Docks to decline to a point at which half are now closed. Only recently has attention turned to the potential of these inner city sites and so far only piecemeal developments have taken place.

At the time that developments were taking place at Tilbury plans were begun for a port to rival Rotterdam further downstream at Maplin, but economic changes brought these to a halt, and the opportunities have been lost, at least for the time being.

I would like to thank Mr R. Glossop for suggesting that I should write this book, and for his helpful support, also Sir Harold Harding, Past President of the Institution of Civil Engineers, who encouraged his young engineers to keep notes—a habit which has proved to be invaluable in compiling the information for this book. Most of all, I thank my wife Marjorie who having lived with engineering for many years, encouraged me in my writing, helped with the research and typed the draft manuscript.

I would also like to acknowledge with thanks the help I received from the following individuals and organizations:

J. R. Presland, BSc (Econ.), FCA, IPFA, MBCS, Vice Chairman and Executive Director of the Port of London Authority, for permission to publish information and photographs from the PLA Records;

J. W. G. Petrie, BSc, FICE, Chief Engineer (PLA) for reading the manuscript and checking the accuracy of information contained in this book;

members of the Chief Engineer's Staff for the enthusiastic help given in assembling photographs of engineering works from time to time carried out in the Port of London;

Sir Edgar C. Beck, MBE, MA, FICE, Chairman of John Mowlem and Company Limited until his retirement in 1979, for permission to publish information and photographs from the Mowlem Records; and

the Council of the Institution of Civil Engineers for permission to quote extracts from papers on the construction of London Docks presented to the Institution.

Contents

List of illustrations

The first hundred years

EARLY HISTORY

Maritime trade has been conducted on the River Thames and in the Port of London since early times. For many years ships berthed at wharves in the Pool of London and discharged their cargo at Legal Quays close to the Customs House. The London Bridge of Peter de Colechurch which lasted from 1176 to 1831 proved to be a great barrier to shipping as only those boats small enough to navigate the difficult ship passage could proceed upstream to small drawdocks excavated in the banks of the river between London Bridge and Blackfriars Bridge.

Below London Bridge a number of small docks were built. Most have disappeared but records have remained of three. In 1661 a small fitting-out basin of 1½ acres fitted with gates to retain the water was built at Blackwall, and was the first wet dock constructed in London. A much larger scheme built in 1703 was the Howland Great Wet Dock at Rotherhithe, extending over an area of about 12 acres. It was entered through a lock 150 ft long by 44 ft wide, with a depth of 17 ft at High Water Spring Tides, and was able to accommodate the largest ships of the time. Because of difficulties with the Customs it was never used to discharge cargo, but only as a sheltered ship repair depot. In 1763 it was sold to a local ship builder and used for the whaling trade, being renamed Greenland Dock. In 1807 it was incorporated into Surrey Commercial Dock.

In 1789 the East India Company constructed two basins at Blackwall, known as Brunswick Dock, used for the repair and fitting out of the Company's ships. It was provided with a large masthouse about 120 ft high, used for stepping masts.

The Admiralty had also carried out developments on the south bank, particularly at the time of the Napoleonic Wars. At Deptford the Royal Victualling Yard was established to service the fleet. Very conveniently on the opposite bank, Lady Hamilton had her residence at Isle House (close to the Blackwall Entrance to West India Dock) so that she could watch Nelson's ship return to base. At Greenwich the Royal Naval College was established, and at nearby Woolwich the Arsenal was built, chosen because the Thanet sand found on the site was a most excellent moulding material for casting gun barrels. The desolate marshes were also ideal for

PLA

West India Dock

firing ranges and testing grounds. Further down the river at Sheerness on the Isle of Sheppey a great naval dockyard was established.

Apart from naval activity, shipping traffic in the port expanded greatly during the 18th century and reached such proportions that there was serious congestion in the river. This led to intolerable theft and organized pilferage. By 1798 ships were arriving in convoy at the Legal Quays between July and October, the season with prevailing winds from the West Indies. As many as 40 000 hogsheads of sugar, piled eight high, would be unloaded at quays only able to accommodate 30 000, coffee and rum being left in the open.

The 1500 ft long Legal Quays had been built in the 15th century; certain other quays 3700 ft long, mainly on the south bank and known as Sufferance Quays, were allowed to be used under licence. No other expansion had taken place. A movement for reform was vigorously pursued by William Vaughan, a London merchant. He suggested that docks should be sited at St Katharine's Church, at Wapping, on the Isle of Dogs and at Rotherhithe, and that these should be developed over the following 30 years.

Opposition to the schemes was led by Edward Ogle, a wharf owner who wanted to dredge a channel from Limehouse to London Bridge, with mooring chains on both banks. There was also a strong lobby in favour of forming a harbour above London Bridge with Legal Quays as far as Blackfriars Bridge. Thomas Telford (the first President of the Institution of Civil Engineers) suggested the complete removal of the existing London Bridge which had been such a restriction to navigation and designed a new bridge with a single cast iron span of 600 ft capable of allow-

2

ing ships of 200 tons to pass below. As the custody of the Thames had been vested in the City Corporation of London by Richard Coeur de Lion since the 12th century, the authorities were in favour of keeping the port within the City boundaries.

Eventually an Act of Parliament was passed in 1799 for the construction of the West India Dock in the Isle of Dogs by the West India Merchants. Whether development should be of docks or riverside wharves was a question raised again in 1905 when the decision was in favour of enclosed docks. However, later developments have returned to riverside jetties and wharves, particularly for oil companies.

Throughout this book there are references to Trinity High Water (THW). In 1800 Captain Huddart surveyed the tidal reaches of the Thames in an endeavour to find a common basis for the establishment of the level of high water ordinary spring tides which he called Trinity High Water. This datum mark was later transferred to the Hermitage entrance to London Docks.

Various other marks were established from time to time purporting to be the same level but, owing to errors in levelling and settlement of the ground, discrepancies of a few inches have arisen. Trinity High Water is now established by statute to be 11.40 ft above Newlyn Datum.

The tidal range in the Thames at London is considerable, about 21 ft at spring tides, and this was one of the factors which led to the establishment of enclosed docks where lock gates could be used to keep the water level constant in relation to the quays.

BUILDING WEST INDIA DOCK

It was in the Port of London that the building of new docks with impounded water and high security fences started. The new methods of handling goods must have had an impact on the wharfingers [wharf-owners] who ferried the goods from moored ships as dramatic as the Container Revolution 170 years later.

It is necessary to follow the development of the odd collection of docks known as the Port of London, if the problems handed down to the twentieth century are to be understood. Three of the new docks built in the second half of the 19th century were a result of envy and cut-throat competition rather than commercial need. The enmity between London and West India docks was bitter and as will be seen finally led to their downfall.

As already stated, in 1799 an Act of Parliament had been passed granting permission to build an enclosed dock and warehouses across the loop in the River Thames at the Isle of Dogs. The new dock company was to have exclusive rights to carry on all trade with the West Indies for 21 years. As the dock was to be sited well away from the city in open country, a main feature of the development was to be the high security walls surrounding it.

William Jessop was the Engineer and John Rennie was called in as special consultant. The first quays were open to shipping in 1803, and the massive warehouses, still to be seen at the entrance to West India Dock, opened in 1806. The dock company, although autocratic, was well run on military lines. All ships were

3

loaded and unloaded by their own labour, and no carmen or outside porters were allowed inside the customs walls. The entrance was patrolled by armed guards. The guard room, or roundhouse as it is now known, is situated near the Police Office.

Trade was good and profits were well up to the 10% expected. The West India Dock Company was happy to be left alone sheltering under the cloak of its monopoly.

Warehouse at London Dock

LONDON DOCK

Nevertheless, right from the start they were not to have things their own way. A year later, in May 1800, a second company to be known as the London Dock Company had a bill passed by Parliament authorizing the construction of a dock at Wapping.

It was nearer to the heart of London and like West India Dock was granted a monopoly to handle tobacco, wine, rice, brandy and other precious goods from all parts of the world except those covered by their rivals' agreement. Again their monopoly would last for 21 years.

John Rennie was appointed Engineer, and the designs of the docks and warehouses were similar to those in the West India Dock. Heavy expense in buying up house property and delays in site clearance postponed the laying of the foundation stone until 26 June 1802. The quays were opened to shipping in 1805 and the warehouses completed soon afterwards.

The Directors of the London Dock Company could not bear to see the rival construction of West India Dock going on apace, and did all they could to hinder progress. It so happened that the brickworks situated in Stepney was nearer to Wapping than to the Isle of Dogs. The price quoted to both developers was the same so it did not take much to bribe the haulier to deliver bricks to London Dock instead of the longer distance to West India Dock, but in spite of being starved of brick supplies, West India scored a moral victory over the new enemy by opening two years earlier. The trade in brandy, French and Spanish wines, to which London Dock had looked for its profit, was badly hit by the Napoleonic Wars and only paid 3% for the years 1817 to 1819.

EAST INDIA DOCK

The East India Company was founded in 1600 and traded exclusively with India, China, and the East Indies. The members already owned land and a shipyard at Blackwall. In 1782 they had built a fortress-like warehouse at Cutler Street in the City, to store the valuables imported from the Far East. Their goods had been handled over open quays, but seeing the other two companies starting on expensive ventures to protect cargo, in 1803 they decided on similar constructions.

Ralph Walker was the Engineer, and John Rennie who had also been retained as specialist consultant on the earlier docks was responsible for the entrance lock.

At that time, as the Commercial and East India Dock Roads had not been constructed, access to the brickworks at Stepney was almost impossible although this was only a few miles away. As it seemed most unlikely that there would be sufficient bricks available for the new dock, Ralph Walker decided that the bricks would be made on site. He let a contract to a brickmaker, who because there were no lodgings available for his men decided to build his own labour camp on a piece of ground nearby. The blue clay made excellent bricks for the dock walls, and part of the excavation was obtained for nothing.

Little foresight was used in choosing the size of entrance locks. The quays were barely adequate for the sailing ships of the time and allowed no room for the larger ships of a few years later.

When in 1833 the East India Company ceased trading, the West India Dock Company bought the dock and some warehouses in Fenchurch Street and Crutched Friars. The large Cutler Street warehouse was sold to the newly formed St Katharine Dock Company.

Because of its small size, the East India Dock played no part in the development of the Port. After World War II it was used partly for coasters trading with the Channel Islands, and partly for storage and servicing of dredgers and marine

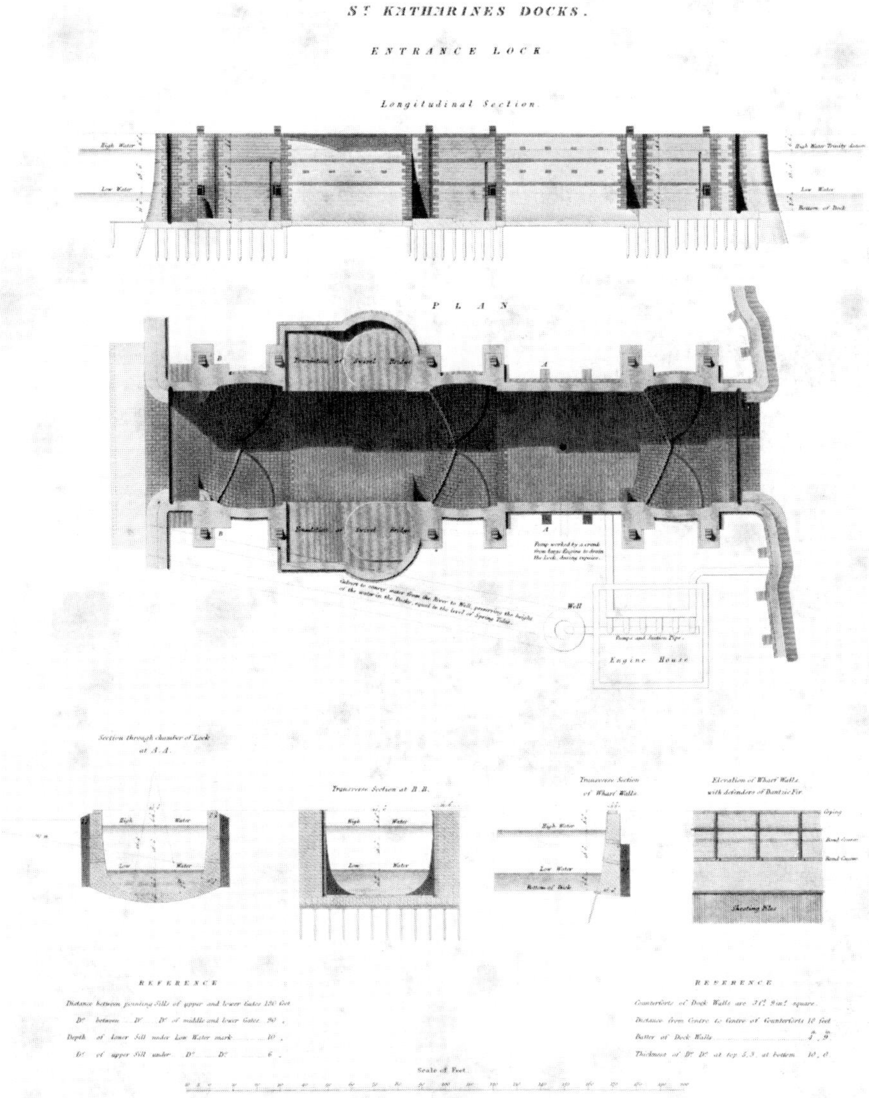

St KATHARINES DOCKS.

ENTRANCE LOCK

Longitudinal Section.

PLAN

Design of St Katharine Entrance Lock

equipment. The export dock housed the Brunswick Wharf Power Station. During that war its main claim to fame was its use for building some of the Phoenix Units for the Mulberry Harbour.

Shortly before the West India Dock Company's monopoly expired, they thought it prudent to apply for an extension of their privileges to handle all West Indian goods exclusively. This was turned down by a select committee of the House of Commons who reported 'that it was not expedient to grant the prayer of the petitioners, but, on the contrary, recommend that, the expiration of those privileges would open the way to free competition among the docks'. The cancellation was to take place in 1827 when all the Acts conveying exclusive privileges would have expired.

ST KATHARINE DOCK

Immediately the news leaked out that privileges would be stopped, influential merchants in the city formed themselves into a Company to develop the land between the Tower and London Dock, constructing a wet dock near to the heart of the city to be called the St Katharine Dock.

The land chosen was 25 acres in extent, and included about 1500 houses, warehouses and sheds, together with a church, houses and grounds belonging to the Hospital of St Katharine. This religious establishment had been founded in 1148 by Matilda, wife of King Stephen, and the church was built in 1443. Although about 11 000 people lived in the area and would have to be rehoused, only a few ever attended church, yet a great outcry was raised against the desecration of the sacred structures, carefully orchestrated by the London Dock Company who did not want to see a competitor on their doorstep. However, an Act enabling construction was passed in June 1825 and the Dock opened in October 1828.

Thomas Telford was the Engineer and Phillip Hardwick the architect for the warehouses and St Katharine Dock House, the imposing headquarters of the Company. The contractors were Messrs Bennett and Hunt, and the work was carried out with such vigour that it was completed in two and a half years. This in spite of a failure of the river cofferdam in October 1827, and a strike the following month.

The main features of the development were the high warehouses flush with the water's edge. Half the ground floor was left open as quay space, the upper floors being supported by massive columns. The scheme was to enable goods to be handled directly from a ship's hold to the warehouse floor by wall crane, thus saving much labour.

The system only worked if cargoes of the same type were stored in the same warehouse. Otherwise sorting normally done on the quay would have to be done at greater cost on the various floors of the warehouse. In later years the problem was less serious. As the entrance lock would not take large ships, the sorting was carried out at the lower docks. Goods were then taken by barge up to St Katharine Dock for warehousing. There was over 1¼ million square feet of storage space, and the warehouses were considered to be some of the finest in London.

The stonework in the entrance lock (and also used in the impounding culverts) was of Bramley Fall, a hard sandstone brought round the coast from Yorkshire. The quality of workmanship was such that practically no maintenance was required until after World War II. The original impounding pumps for maintaining the

J. Flowerday

Detail of St Katharine Entrance Dock

water level in the dock were supplied by Boulton and Watt, James Watt being the son of the engineer of steam engine fame. In 1856 Sir William Armstrong, inventor of the hydraulic accumulator, built one of the first hydraulic pumping stations on the site of the Boulton and Watt pump house, and hydraulic machinery was installed to work the wall cranes used to lift goods to the various warehouse floors. The opening and closing of the dock gates was not changed to hydraulic power, however, and hand winches and chains were retained until 1957.

The layout of the two docks with their sharp angles and connecting basin was designed to give the maximum length of quay in the limited space available. However, it was difficult to warp sailing ships into the awkwardly shaped berths, because, as in the case of the East India Dock, little thought had been given to the future development of ships. The entrance lock, 181 ft long, 45 ft wide, and 25 ft deep, would only take the sailing ships of the time. In consequence in later years the dock could only be served by coasters of up to 1000 tons capacity and by lighters ferrying goods from the lower docks. With too much warehousing and too little quay space the dock company was never really viable. The best dividend paid to the shareholders was 5%, the usual being between 2½% and 3%. The original capital of £1 352 800 had been increased to £2 632 000 and included loans for the purchase of the nearby river quay Iron Gate Wharf, and the Cutler Street warehouses. With the ever-increasing competition between the docks the returns were insufficient to finance the loans, and in 1864 it was agreed to amalgamate with the London Dock Company.

During the period of the monopolies London and West India docks were well organized, and above all profitable. With the large capacity of additional warehousing now available at St Katherine Dock, London had too many warehouses with insufficient goods to fill them, with corresponding loss in revenue. This was the time seriously to consider amalgamation and to stop competition for the available trade, particularly as steam ships could now arrive regularly and were not delayed by northeast winds as was the case for sail.

However, instead of amalgamating competition increased with the Victoria dock to be built on the downstream side of Bow Creek near East India Dock.

VICTORIA DOCK

As far back as 1842–43 Mr Blyth, the manager of the West India Dock Company, had mentioned to Mr G. P. Bidder, the experienced Victorian railway engineer, that the land between Bow Creek and Gallions Reach would be ideal for future dock construction. Having made the suggestion and pointed out the advantages, he took no further interest in the project but was content to concentrate his interests in West India Dock.

Bidder, who at the time was engineering the Eastern Counties Railway from London to Southend, was strongly impressed with the idea. Thomas Brassey, the 'Railway King', was the contractor for the line and the two men joined with another contractor, Sir Samuel Morton Peto and with his brother-in-law Edward Betts in privately financing the construction of a branch line from Stratford, East London, to end in a field at North Woolwich. The line was fondly known as 'Bidders Folly'. In the meantime they quietly started buying up land in the area, some for as little as £6–£7 per acre, although one parcel of 647 acres belonging to the Dean and Chapter of Westminster was to cost dearly. The Dean, hearing a rumour that a dock was to be built, held out for £250 per acre. Over £90 000 of hard-earned contractor's money found its way into the landowner's pocket. It is a matter for speculation how much of this profit reached the Abbey funds.

Built between 1853 and 1858, the new venture was to be known as Victoria Dock. It was 4050 ft long and equipped with four jetties each 581 ft long and 140 ft wide providing nearly 3 miles of quays. The entrance lock, 80 ft wide, 326 ft long and 28 ft deep, was capable of taking the largest steam ships envisaged at the time. The Victoria Dock was considered to be the best dock in the world. The connection with the Eastern Counties Railway put it in direct communication with Shoreditch and Fenchurch Street Station. Warehouses were built nearby, and connections with the Great Northern Railway gave direct access to the rapidly growing Midlands manufacturing towns.

Mowlem

Western Entrance, Victoria Dock

The dock was equipped with the latest hydraulic machinery to put it considerably ahead of its rivals. For instance the lock gates could be opened in 1½ minutes instead of the 10 or even 20 minutes that the operation took elsewhere. Hydraulic capstans operated by a single man saved thirty or forty men hauling on a windlass. Hydraulic cranes were installed on the quays. To give an idea of the success of this new dock, during the month of April 1858 2500 barges and 508 ships entered the dock.

As can be imagined, the impact on West India Dock was far reaching, particularly as the entrance lock there was only 45 ft wide, and the owners could not afford to rebuild it. The initiative for obtaining the new steam traffic was left with Victoria Dock.

10

The whole enterprise had been a contractor's speculation costing £1 076 664, and although extremely successful, by the time of the financial crash of 1866 there was still nearly £800 000 outstanding on the loan. The crash was brought about as a result of the Railway mania of the 1840s. Investors invested, finance houses made advances, engineers planned, and contractors built, regardless of the ultimate profitability of the ventures. The large financial houses of Overend and Gurney lent money recklessly without sound security. After the Indian cotton collapse of 1865 they ran short of liquidity. Their liabilities were massive at £18 727 917 (about £500 million at 1979 rates).

Failures and bankruptcies snowballed, and on the 11 May 1866 the bank rate was raised to the almost unprecedented level of 10% in an endeavour to stop the crash. Credit became unobtainable.

Samuel Peto, who was thought to be one of the most sound contractors, was amongst the first to go, leaving Thomas Brassey with the liability for the whole loan. Brassey was in financial difficulties at the time with a railway contract in Denmark, and had not thought it necessary to limit his private liability. He was unable to obtain more than £30 000 credit, so regretfully was compelled to sell the Victoria Dock to the London and St Katharine Dock Company.

This left West India Dock sandwiched between two docks belonging to London Dock Company, something which could have been avoided with better sense in 1843. A second opportunity came to amalgamate, but again was passed over, owing to stubborn pride; instead the battle continued for a further 25 years. It seems that those who spend their lives inside the high custom walls live a blinkered existence. When in 1864 a further new dock was projected on their very doorstep, it is unbelievable, but a fact, that the West India directors did nothing to prevent it.

MILLWALL DOCK

The land forming the undeveloped part of the Isle of Dogs, some 200 acres in extent, was bought up by agents at agricultural rates. The Millwall Freehold Land and Dock Company was formed and a Bill enabling the construction of a new dock was rushed through Parliament before the West India Dock Company woke up to the danger.

Work started on the project in June 1865 and was finished three years later. John Fowler was the Engineer and Sir John Aird the contractor. The new dock was L-shaped with an 80 ft wide entrance lock into the river, and 36 acres of water with quays each side through the Isle of Dogs right up to the boundary with West India Dock. Features of the project were the first granary built in the Port of London for the Baltic grain market, and the first dry dock in an impounded dock. The sheds along the quayside were two storeys high with the low headroom of 9 ft 0 in. to take hogsheads holding 52½ gallons stored two high. Larger warehouses were provided on the west side for longer term storage, though again with limited headroom.

The dock was not a commercial success. In order to attract customers, the rates were set too low to be profitable. To make matters worse the rates were based on a level of traffic which was never attained, and consequently the losses increased. Amalgamation with West India dock took place, but the competitive rates had done nothing but harm both to the new company, and to their neighbours who had joined in the senseless price cutting.

In 1870 Colonel J. L. Du Platt Taylor, who had been newly appointed Chairman of the West India Dock Company, made attempts to bring his board of Directors up

K shed, South Quay, West India Dock

to date with developments. He called for modernization of the seventy-year old dock. Although needs were changing it was decided to build a new south quay with warehouses, in spite of the fact that it was becoming important to get goods from ships to customer quickly by using transit sheds. No alteration was made to the narrow entrance lock, so the feeble attempts at modernizing the dock brought no new traffic and wasted money which could be ill afforded.

ALBERT DOCK

Having seen the West India Dock Company embark on the south quay expansion scheme, the London Dock Company decided to increase their stranglehold by adopting the plan which Bidder had originally proposed of joining Victoria Dock to Gallions Reach by a new dock to the east of it.

In 1874 construction commenced on the Albert Dock scheme. Thomas Brassey had died in 1870 and the contract was awarded to Lucas and Aird who had just finished the construction of Millwall Dock. Sir Alexander Rendel was the Engineer. When the dock was completed in 1880, there were 175 acres of water and 7 miles of quays with a depth of 30 ft when combined with Victoria Dock. Two entrance locks connected the system to the river through a basin east of Manor Way. The new dock was opened by the Duke of Connaught, and Queen Victoria gave permission for the group to be called the Royal Victoria and Albert Docks, later known as the 'Royals'.

Lit by electricity, supplied with hydraulic cranes and steam winches, it was able to take the largest ships afloat (up to 12 000 tons) and turn them around more quickly than any other dock. A sophisticated system of railway sidings gave direct access from the transit sheds along the quays to the Great Eastern Railway. A new venture was the Gallions Hotel which provided for the needs of passengers and their friends coming from Fenchurch Street by boat train.

It was considered to be the finest dock in the world, and effectively took away the small amount of steam shipping which could get into West India Dock.

DECISION TO GO TO TILBURY

Du Platt Taylor and after him the new Chairman Harry Dobee, seeing that the West India Dock faced ruin, refused to give in and amalgamate with the enemy. Instead they decided to risk the company's depleted finances in one last daring scheme. They would go 26 miles down the river to Tilbury and build a new dock there. The site chosen was a deserted piece of the Essex marshes opposite Gravesend.

They argued that the journey to London could be cut by at least a day, saving pilot and towage charges. They envisaged all the big ships turning into the new dock instead of proceeding up to the Royals. The promotion was not based purely on commercial considerations but tinged with envy of the London Dock Company who in the Royals possessed the finest dock in the world, filled with the latest steam liners. They were so confident of the success of their brainchild that their enthusiasm knew no bounds and their capacity for spending money grew with their enthusiasm.

Having learned from the promotors of Millwall Dock the secret of buying land before the project was made public, they appointed agents to buy 450 acres of marsh land at agricultural rates.

Quay transit sheds, each with rail sidings connected to the Eastern Counties railway, and a large warehouse in Commercial Road close to Aldgate would ensure rapid transit of the goods. So that unloading could be carried out in the dry, the rail trucks were to be shunted right inside the warehouse for distribution or storage of the cargo.

To ensure the comfort of expected passengers and their friends a first class hotel was built with a commanding view of the estuary and North Downs beyond.

Accommodation was provided nearby at Tilbury Gardens for the officers. Houses were built in Orient and Peninsular Roads for the staff and tenement blocks for the dockers. It was a bold plan, and on paper a winner!

A Bill was rushed through Parliament and such was the enthusiasm to get started on the new project that the first sod was ceremoniously cut on 17 July 1882, just two weeks later.

A contract worth £1 100 000 had been let to the Contractors, Kirk and Randall. Almost immediately they were in trouble trying to excavate the soft blue estuarine clay and peat in an endeavour to found the dock wall in sand and gravel some 45 ft below. No major civil engineering work had been carried out in this area before, and the scope of the work was quite beyond the contractors' ability. The promotors

PLA

Excavation for Branch Dock, Tilbury, showing peat

were impatient to see rapid progress, and in fact had paid good money in extras to make up for time lost, but all to no avail. By 1884 the contractor was thrown off the site and Lucas and Aird, who had worked on the Millwall and Albert docks, were appointed to finish the works. This was done in two years and the dock opened in April 1886, but at a cost of twice the original estimate. The success of Lucas and Aird's work was brought about by the draining of the underlying gravel by the use of many sumps, some up to 70 ft deep. This not only allowed the soft clay and peat to dry out but drained the water from village wells for miles around.

In the meantime the London Dock Company, seeing the activity 'down river' embarked on a quiet programme of sabotage. They managed to persuade consignees of cargo to qualify their bills of lading with the clause that no goods should be unloaded at Tilbury. The lightermen were also persuaded to black the new venture.

After the first ship had entered the dock for the opening ceremony no others followed it. In fact the sabotage had been so effective that the dock remained unused for four months. Finally as a desperate measure the Clan Line was bribed to leave the Royal Docks by an agreement that it should pay only half the rates for 10 years, and 80 years later they were still using Tilbury as their base.

At Tilbury the tidal basin entrance locks and the impounding station were all sited on the inside of a bend in the river, in slack heavily silted water. Not only did the tidal basin silt up rapidly, but the silty water was transferred to the dock through the impounding pumps. This necessitated dredging inside the dock as well as regular dredging in the basin. (In fact before the new impounding station was built in 1962 the Port Authority was spending £750 000 per year in dredging the dock.)

When the London Dock Company saw trade being seduced away from the Royals, they were furious, and immediately cut their rates unmercifully. The West India Dock Company, beset by disputes and litigation with the contractor who had been turned off the site, was forced into liquidation only two years after the grand venture had opened to shipping. The gamble, brought about by greed and envy, had lost everything.

On 1 January 1889, a working arrangement was reached between the two rivals, 64 years of bitter feuding too late. Price cutting and excessive competition had destroyed the docks. An amalgamation, to be known as the London and India Docks Joint Committee, was set up and lasted until the formation of the Port of London Authority in 1908.

DEVELOPMENTS ON THE SURREY SIDE

No mention has been made so far of the Surrey docks because they took no part in the battles between the dock companies on the north side of the Thames. Communication by road was difficult, the distance from the India Dock via London Bridge (the nearest crossing point on the river) being about seven miles.

The first schemes for the development on the south bank were by Ralph Dodd, a well-known promotor of the 18th century. At the height of the canal age he thought up a grandiose scheme for building a canal from Rotherhithe through Peckham to Kingston and Epsom, with an extension to be taken through to Portsmouth and Southampton. This would enable warships to sail from Chatham Dockyard to Portsmouth without attracting the attention of the French Navy in the Straits of Dover during the Napoleonic Wars. He thought that the Admiralty might finance the project. In spite of lack of interest by the Navy, the Grand Surrey Canal Company was incorporated in 1801 and opened in 1807. An entrance basin and lock were built at Rotherhithe and the canal was completed as far as Peckham with a branch to Camberwell before funds ran out.

Construction of Tilbury Entrance Lock and Dry Docks, taken from balcony, Tilbury Hotel

The scheme was never completed to Kingston, as the railway era made canals no longer viable. The Canal, however, hummed with industry, small businesses and timber wharves being established along its bank. As late as 1950 one of the unusual trades being carried on in an old corrugated iron shed was the making of 'penny dabs' of whiting for local housewives to use on their front door steps. The advent of the tower block killed the trade, and the few remaining front door steps are no longer whitened.

In 1807 the Commercial Dock Company was formed to develop the old Howland Great Wet Dock or Greenland Dock. Two more companies emerged in 1809 to exploit the timber trade, the East Surrey Dock Company, and the Baltic Dock Company. These built Lady Dock, Lavender and Acorn ponds and Quebec and Russia Yards.

The companies all appeared to work amicably together and unlike the rivals on the other side of the river sensibly saw the advantage of amalgamating into one business. This took place in 1850, under the name of the Surrey Commercial Dock Company. Trade was limited to the importation of timber and grain but the business was a commercial success. The absence of feuds allowed the directors to culti-vate the needs of their customers.

The only new work undertaken was the opening of Canada Dock in 1876. In 1895, however, the Company decided to expand and a difficult undertaking was embarked upon in the construction of a widened Greenland Dock and a new entrance lock of modern proportions. Sir John Wolfe-Barry was the Engineer, and like so many engineers after him fell foul of the fine-grained Thanet sand under-lying the clays of the Woolwich and Reading beds. The new complex took nine years to construct, not being opened until 1904.

The scheme had been too costly at nearly £1 million and did not really attract new trade. To commission the dock some shipping was bribed to leave the Royal Docks. The London Dock Company, as might have been expected, offered cut prices to the timber trade of '25% off', but the only effect of this latest activity was to shorten the time taken to bring to an end the private dock companies and establish the Port of London Authority in 1908.

It is interesting to see 70 years later that London, St Katharine and Surrey docks are all closed. Although only a few ships are now using the Royal Docks, Tilbury and to some extent India and Millwall remain busy. London Dock, the cause of so much trouble, is being redeveloped as a housing estate!

The Port of London Authority, 1908–1939

FORMATION OF THE PLA

The London docks complex grew up in the 19th century to no preconceived plan. The siting of some docks had been on the principle of 'How can we catch ships coming up the river before they get to our competitors?' rather than by considering the best locations to make London the greatest port in the world.

The continuing bitterness between dock companies only underlined the fact that they were exhausting their financial ability to modernize the old installations. During the last few years of the Victorian era, the British Empire was expanding rapidly, and to meet the demands of increasing trade, ships were getting larger at an alarming rate. Apart from the docks becoming run-down, there was also great concern about the state of the river. Shoals of ballast had obstructed the channel, which itself was insufficient to meet the needs of shipping. It became evident that the Thames Conservancy, who were responsible for dredging the tideway at that time, were unable to improve the channel sufficiently without drastic changes in their methods of working. The dock companies could gain no advantage by deepening their entrance locks if the river remained unimproved. There was only 16 ft depth of water above Gravesend, and 24 ft at low water spring tides below that point.

The cost of deepening the channel was estimated to be about £2.5 million – quite outside the financial capabilities of the Thames Conservancy or the private Dock Companies.

King Edward VII's speech at the opening of the new Parliamentary Session of 1903 included a bill for the establishment of a new Port of London Authority. However, no Parliamentary time could be found in that session to discuss the matter. During 1905 and 1906 opinion was divided between expanding the riverside wharves and quays, or the docks. Lloyd George, President of the Board of Trade, visited continental ports to settle the matter, and came down strongly on the side of retaining the docks. During this time the private dock companies, seeing that they were likely to be amalgamated into a public authority, for the first time agreed to work together. They increased the dock dues considerably in order to be

18

in a strong position when negotiating compensation claims, and put forward their own further proposals for working together. The proposals had come too late, and in any case would not have solved the problems of deepening the river. Lloyd George finally agreed the compensation terms and presented a Bill to Parliament in April 1908. Winston Churchill had become President of the Board of Trade by this time and it was he who saw the Bill become law in December.

The Port of London Act (1908) created a self-supporting public body to administer, preserve, and improve the port, ploughing any profits back into the undertaking. Capital was to be raised on the open market by the issue of 'Port Stock' in limited amounts at fixed rates of interest. Revenue was to come from statutory dues and rates on vessels and goods, and from charges made for services and accommodation.

The new Authority was to inherit the freeholds of the five enclosed docks, and was to take over the responsibility of Conservator for the tidal reaches of the Thames from Teddington Lock downstream to the seaward limit at the Nore.

The Board of ten appointed and eighteen elected members was rather large and varied in composition by modern standards, being composed of the Mayors of the six Boroughs in which the docks were situated, representatives from the London County Council (as it then was), the City of London, the Board of Trade, the Admiralty, shipping companies, Trinity House, stevedoring companies, masterlightermen and barge owners, tug owners, a Chairman and a General Manager. Each had his own sectional interests to put forward, and it was only through the integrity and personal drive of the Chairman that the Board was welded together into a workable unit.

FIRST CHAIRMAN AND CHIEF ENGINEER

Sir Hudson Kearley (MP for Devonport) was appointed Chairman. He had been a Director of Kearley and Tonge, and the Independent Tea Stores. He was created Lord Devonport in 1910. During World War I, apart from being Chairman of the PLA, he was to become the Nation's first Food Controller (for which he was created a Viscount).

He was an autocrat of enormous vision and driving force. He had such awe-inspiring presence that 30 years after he had retired from the PLA in 1925, Sir George Burt (who knew him well) used to talk of his meetings with Lord Devonport.

The first Board meeting was held on 16 March 1909, and priorities were decided: repairs and improvements to the existing dock premises which owing to neglect were neither efficient nor safe; the extension and deepening of the existing channel in the river; the building of a new Head Office where the scattered administration of the former dock companies could be brought together and co-ordinated, and the construction of a new dock at North Woolwich, later to be called King George V Dock, with a new large entrance lock into the river, and connected by a cutting to the Victoria and Albert Docks. The total cost of these engineering works was estimated to be £14 million to be spread over ten years.

To put these plans into action Mr Frederick Palmer (later Sir Frederick, KCMG) was appointed to the post of Chief Engineer. Like Lord Devonport he was a man of vision who was determined not to be trapped into building on a small scale as he foresaw larger ships as inevitable. His son, the Consulting Engineer John Palmer, speaking at the Institution of Civil Engineers on 10 March 1953 said of his father Sir Frederick 'I had at one time been rather well acquainted with that gentleman, and in my opinion I believe that his main contribution to the greatest Port in the world had been to produce in 1910, a scheme for nine major capital improvements spread over the 30 miles length of the River'. If it had not been for the materialization of the ideals of that particular Engineer, there would be today no King George V Dock, and no dry dock or passenger landing stage at Tilbury.

It was Palmer's view that what the Port of London needed, if it was to be the greatest port in the world, was plenty of idealism from engineers, and not too much commercial expediency. One of the big problems facing him was to judge the likely size of ships 10 or 20 years in the future. The cost of capital works he thought would vary approximately with the cube of the draft of the deepest ship for which the port was intended to cater. The advantages of building bigger ships accrued much more to the ship owners and very little to the Port Authority. Nevertheless, it was important to build on a scale sufficiently large for the entrance locks to remain unaltered for many years without modification or rebuilding.

In spite of his far-sighted ideals of building 'big', financial restrictions were imposed and often cuts in expenditure were made without realizing the troubles being built up for the future. The two most important entrance locks were reduced in size from the proposals of 1910. The recommended length for the Tilbury new entrance lock had been 1050 ft, with a width of 130 ft. It was actually built 994 ft long and 110 ft wide. The recommended length for the King George V entrance lock had been 850 ft and 110 ft wide – it was built 800 ft long and 100 ft wide.

Tilbury entrance lock was capable of accepting all large ships including specially designed container ships of the first generation, such as *OCL Jervis Bay*, some 65 years after it was first planned in 1910. In the same way the King George V entrance was able to accept all shipping intending to use the Royal Docks. However, the Dockmaster had cause to wish that Sir Frederick Palmer's dimensions had not been reduced when in 1939 the Cunard liner *Mauretania* of 35 739 tons entered the lock with only six feet to spare each side. On being appointed Chief Engineer, Sir Frederick Palmer saw his two most urgent priorities to be the creation of a deep dredged channel in the river, and the modernization and extension of the Royal Docks.

DREDGING THE THAMES

Until the end of 1857 dredging of the navigable channel had been carried out for the City of London Corporation by 'ballast men'. These removed the extensive shoals of gravel which affected navigation. After the Dagenham breach of 1707 the Lord Mayor reported that although many thousands of tons of ballast had been taken up annually the shoals increased daily.

In 1858 the Thames Conservancy took over the responsibility for the tideway, and proposed a dredged channel 200 ft wide below London Bridge, widening out to 600 ft at Crayfordness some 16 miles downstream. The depth would be 18 ft at London Bridge gradually deepening to 24 ft near the estuary.

As shipping increased, complaints became frequent. In 1902 the Board of Trade set up a Commission under Sir John Wolfe Barry, who reported that the Thames Conservancy plans were not enough. As a result an Act of Parliament in 1905 empowered the formation of a channel 1000 ft wide and 30 ft deep at mean low water spring tides.

In the 31 years from 1878 until the responsibility for dredging was taken over by the Port of London Authority on 1 April 1909, the Thames Conservancy removed 11 144 156 cu. yd measured in hoppers, or nearly six million cubic yards solid. This resulted in a considerable improvement to the tideway. It did not meet the requirements of the larger ships, however, and trade began to drift away to the Continent and other UK ports.

To stop the further loss of business Sir Frederick immediately ordered a fleet of large bucket dredgers and hopper barges to add to the plant taken over from the

Excavation for King George V Dock

PLA

Thames Conservancy. In many parts of the river it was necessary to realign the dredged channel to prevent it silting up and to obtain the maximum scour.

The dredging plan adopted was slightly less ambitious than the one outlined in Sir John Wolfe Barry's report. From the seaward limit to Coldharbour point, a channel 1000 ft wide and 30 ft deep at mean low water spring tides was dredged. From that point up to Albert Dock it was 27 ft deep and then up to West India Dock it was maintained at 20 ft deep.

Dredging was carried out continuously for 15 years before the new channel reached completion. All dredged material was taken out to sea and dumped in Black Deep 30 miles seaward of Southend. It was not until 1963 that a survey using radioactive isotope tracers showed that much of the material dumped in Black Deep found its way back up the river by the action of the tides. Arrangements were then made for dredged materials to be pumped ashore at Rainham in Essex, to reclaim marshland.

In 1908 the seaward limit of the PLA was moved to a point between Havengore Creek on the Essex shore to Warden Point in the Isle of Sheppey.

So successful was the dredging programme that between 1909 and 1930 the PLA had dredged 41 768 747 cu. yd measured in hopper barges. Somebody calculated that this quantity would make a mountain half the size of Snowdon!

During this period nearly five million cubic yards of ballast had been dredged, and about one million cubic yards of mud were removed annually, to keep the deep water channel clear. This was a rate over $5\frac{1}{2}$ times faster than had been carried out by the Thames Conservancy earlier. Having arranged for the dredging programme to be put in hand, Sir Frederick Palmer embarked upon a decidedly extensive plan for reshaping the port. This he entered into with a sense of public duty within the strict financial limits imposed by statute. Cyril Kirkpatrick (later Sir Cyril) was engaged as Chief Assistant Engineer in 1910 and some three years later followed Sir Frederick as Chief Engineer. He was thus able to see the schemes that they had planned together brought to fruition – Sir Frederick Palmer was retained as a Consultant after he had left the PLA.

Although modernization was to take place in all docks, the two men concentrated on the Victoria and Albert Docks. This was because the fruit trade was expanding enormously with the fast steam ships. The coal trade was also growing with William Cory's improved methods of discharging cargo. Above all, refrigerated foodstuffs, particularly meat and dairy produce, were now being imported from Australia, New Zealand and South America. Timber imports and paper pulp grew with the increasing demand. The export of manufactured goods from the Midlands was expanding, and cement became a big export trade. Oil depots were set up at Thameshaven and large quantities of oil began to come into London. The volume of trade in 1909 was 40 million tons or 29% of the UK total, including the private Riverside Berths.

It was with this upsurge in trade as a background that the extensions to the Victoria and Albert Docks were planned. Two dock expansion schemes were considered, one to the north of the existing Albert Dock, and another to the south (later built) to be named the King George V Dock.

PROPOSED NORTH ALBERT SCHEME

The proposed north Albert Dock scheme was vast in comparison to the existing docks. It was to be sited in the Beckton Marshes running parallel to the existing Albert Dock, with an entrance lock through what is now the Tate and Lyle sports ground, and the Beckton surface water pumping station. It was to have 126 acres of water area with 15 600 ft of quay providing berths for 25 ships. The entrance lock 1000 ft long, 120 ft wide, with a depth of 52 ft below THW, was larger than anyone had built before. At each side of the entrance lock there was to be a dry dock, one large, almost as big as the entrance lock, the other slightly smaller.

This ambitious scheme was considered to be beyond the financial limits imposed on the port, so the Authority favoured a more modest scheme smaller than the North Albert, situated to the south of Albert Dock. Again this would be parallel to the existing dock, with an entrance lock 800 ft long, 100 ft wide but with a depth of only 45 ft below THW. At 63 acres of water it would be three-quarters the size of Albert Dock with berths for 13 ships. At the west end there was to be a dry dock 750 ft long, 100 ft wide and 30 ft deep.

It is interesting to speculate as to the effect the North Albert scheme would have had on the Port of London, had it been built. It would have doubled the size of the Royal Docks, provided deep water berths, and would most certainly have made London the envy of other ports all over the world. During the boom time of the late 1950s when ships were double-banked in the Albert Dock, its additional berths would have been very welcome. It could have become London's container port, with the large areas of the Beckton marshes available for the back-up room necessary for such operation, although the increased depth of the dredged channel to make it viable for container ships would have been costly. In later years other engineers reviewed the scheme, but again thought it was too ambitious.

BUILDING OF KING GEORGE V DOCK

Although outline schemes were drawn up in 1910 for both docks, arguments, and divergent opinions were not settled until the summer of 1911 when the North Albert dock scheme was finally discarded. The King George V dock layout having been decided, detailed designs were started in the autumn of 1911 and were completed in time for a contract to be let in August 1912.

S. Pearson and Son, the subsidiary construction company of Lord Cowdray, were awarded the contract, the works to be completed within four years.

Before the whole of the site could be made available, 204 new houses were built to replace those which had to be demolished at North Woolwich. These rehoused the people on a 'model garden village principle' in Prince Regent Lane, at 22 houses per acre.

Excavation for the dock was carried out with 'steam navvies' loading into 4½ cu.

yd side-tipping wagons, or 10 ton end-tipping wagons carrying about 15 cu. yd of soft clay. About 50 000 cu. yd of material were handled per week using 17 miles of 4 ft 8½ ins track, 16 'six-coupled' steam locomotives and 620 wagons. Many of the locomotives were working in the dock up to the 1960s when they were replaced by diesel locomotives.

With the exception of the Chingford reservoir in 1936, this must have been one of the last major earth-moving operations in London using techniques developed in the railway era.

Over 900 men were employed, many in laying and slewing the tracks to ensure that wagons were within reach of the machines loading them. A large number of derailments which inevitably happened in this class of work were quickly dealt with. Some railway plate-laying gangers took great pride in being able to put a heavy locomotive back on the rails almost singlehanded, with the aid of jacks, packings, and levers.

There were five million cubic yards of excavation, of which two million cubic yards of unsuitable material were taken out to sea in hopper barges. Over one million cubic yards of good quality sand and gravel were recovered from the site and used in the works. About half was required for constructing the dock walls and entrance locks, the remainder being used for filling in behind the walls and as a top layer in the reclaimed areas.

With the outbreak of war in 1914 the contract almost came to a standstill as men of military age joined the armed forces. About 250 men, all over age, continued doing what they were able, but it was found necessary to terminate the contract by agreement and continue by what was called 'direct administration'. In the autumn of 1918 with the war drawing to a close, facilities were granted for labour and materials to be obtained. In retrospect it seems that the construction work was to be used as a convenient 'home' for men returning from the forces who had no immediate job to go to.

From early 1919 to 1920 the labour force was increased to a peak of 1700 men, the direct labour system being used to accommodate them.

By the end of 1919 work had progressed sufficiently to contemplate the problem of filling the new dock with water. As soon as the inner gates to the entrance lock had been assembled in the dry, they were closed, and as a safeguard against failure, an earth dam was built behind them in the body of the lock. The middle gates were to take five months to complete, and the outer gates ten months, so considerable time could be saved by flooding the dock immediately the inner gates were ready.

The water in the Albert Dock was raised to 2 ft 6 in. above THW. A flume 8 ft wide and with its bottom constructed 2 ft below impounded water level connected the northwest corner of the new dock to the water in Albert Dock. On 12 January 1920 Lord Devonport opened the sluice, and the water in the new dock rose by about one foot per day.

Filling was completed after the three sets of dock gates had been tested in August

1920. On 21 February 1921 the PLA Board had a private opening, and viewed the completed works. The official formal opening was carried out by King George V on 8 July 1921.

IMPROVEMENTS TO VICTORIA AND ALBERT DOCKS

Although the building of the King George V Dock was considered to be of major importance, Sir Frederick Palmer insisted on a programme of modernization for the Victoria and Albert Docks, to run concurrently with the new construction. In particular he declared the system of finger jetties in Victoria Dock to be out of date, and preferred straight quays. The jetties had been built in the lightest of construction and 'did not possess that amount of security which was required, and were not of a permanent character'. The Board did not consider the work to be of sufficient urgency so the new construction was not put in hand until 25 years later.

Work which was given the go-ahead was the replacement of the old steam pumps in the Gallions impounding station by new electric pumps in 1911. This enabled the level of water in the docks to be raised to 2 ft 6 in. above THW.

The western dry dock was modified and enlarged to 575 ft long and 80 ft wide in 1914. Even at the height of the war, effort and materials were made available to provide new 3-ton electric cranes and track to the north side of Albert Dock in 1916. This was followed by the layout of 6000 yards of exchange sidings in Victoria Dock near the Custom House Station in 1918.

With the Armistice in 1918 thoughts were turning to the problems of meat supplies coming into the country from Australia, New Zealand, and South America. Number 6 cold store, insulated with two 3-in. thick layers of cork, was built in 1920. At the same time a two-storey reinforced concrete cold store was built at the west end of Albert Dock. This was 123 ft 6 in. wide on the ground and 1100 feet long. It was used as a transit shed, but the upper floor, being only 100 ft wide, provided a balcony upon which meat could be unloaded for cold storing. The upper floor was refrigerated, and a bridge connected it with No. 6 cold store. This scheme had a four million cubic feet capacity and could store nearly a million carcasses of mutton.

This meat-handling scheme was so successful that it was quickly followed by two more in the Victoria Dock. In 1926 a special berth for the Royal Mail Line, importing South American beef, was opened. It was equipped with 6000 feet of mechanical runways on which the meat was hung, travelling over automatic weighing machines before being loaded into insulated road or rail vehicles. About 3800 tons could be handled per week. A similar scheme was constructed two years later for the Blue Star meat ships.

Apart from South American beef, the Victoria Dock captured a large section of the tobacco trade between 1920 and 1922. First a 580 000 cu. ft warehouse was built to take American hogsheads, each weighing half a ton, handled and stored five high using overhead electric cranes. This was followed by the six-storey M warehouse which added a further 1½ million cubic feet of storage space. At any one time there would be between 15 000 and 20 000 tons of tobacco in bond.

The growth of the Mediterranean citrus fruit trade needed special consideration. A coaster calling in various ports in the Mediterranean would pick up quantities of fruit from point to point. On arrival in London it would have to be sorted out to as many as sixty different marks to ensure they were sent to the correct markets. A special shed for oranges was constructed with 40 000 sq. ft of floor space and 500 linear ft of covered platforms for delivery to road or rail vehicles. It was completed in 1935. The boxes of oranges landed on the quay were taken by hand barrow and stacked in the shed according to the coloured mark on the case. Any damaged or overripe fruit could be loaded into a railway wagon for disposal. An unforeseen problem was that juice dripping out of the truck soon ate deep holes in the concrete roadway below.

In 1934 the banana berth was provided to unload Fyffes ships using mechanically operated bucket conveyors which reached into the holds. A discharge rate of up to 80 000 stems per day was achieved with loading into road or rail vehicles. The handling of grain also became a major operation when large silos were established on the south side of Victoria Dock.

The western entrance lock to Victoria Dock had never been very robust, and after the depth of impounded water had been increased, much trouble occurred with

Entrance Lock, King George V Dock

PLA

the gate roller paths. Instead of enlarging the lock to the proportions of 700 ft long, 100 ft wide and 42 ft deep which Sir Frederick Palmer had planned, it was decided to patch it up, and install new gate platforms on top of the old ones. This effectively reduced the depth to the sill to 20 ft 6 in. below THW. This meant that the lock could only be used by barge traffic. At the height of prosperity in 1959 and 1960 when ships and barges jammed the Albert Dock the decision was much regretted as all ship traffic had to use the Gallions and King George V entrances.

On the other hand, once this decision had been made, the way became clear for the building of Silvertown Way, as there could be a fixed span over the entrance.

When the Silvertown Way was opened in 1934 it was hailed as one of the most important commercial highways in the realm, but was only a small drop in the ocean, as road approaches to the port have always been a major problem leading to the devaluing of the upper docks. Years later Tilbury was chosen as the container port rather than the Royals, not only because of its deep water access but because being 26 miles down river from the City, it was further from the urban congestion from which the Royals suffer.

In 1920 the volume of trade using the Royal Docks was just three million tons. In the 10 years to 1930 it increased to eight million tons. With this massive increase, the PLA decided in 1935 to go ahead with a multi-million pound scheme to improve the docks still further.

DEEPENING OF THE ALBERT DOCK AND CONNAUGHT ROAD CUTTING

In spite of the improved depths provided by the new impounding pumps in 1911, larger ships demanded still greater depth. In order to provide this, it was decided to deepen the Albert Dock by 4 ft 6 in. to 34 ft. To prevent the quay walls from being undermined, a false quay was required. Owing to the comparatively narrow width of the dock it was necessary to encroach on the water area as little as possible, and the new quay, 19 ft wide, consisted of a reinforced concrete deck carried on a deep coping beam bearing on cylinders 24 ft apart, these being tied into the existing quay by transverse beams. In order to cause as little inconvenience as possible, work started at both ends of the north quay simultaneously, and by working night and day, using rapid-hardening cement, the 5450 ft long quay was completed in 39 weeks.

When the Albert Dock was first constructed, the North Woolwich line was taken in tunnels under the Connaught Road cutting joining the Albert and Victoria Docks. Sir Frederick Palmer had thought that it would be impossible to deepen this cutting, and that the ships entering Victoria Dock would be limited to 28 ft draught. However, when Asa Binns became Chief Engineer the whole problem was reconsidered. A scheme was devised whereby the tunnels would be lined with rings of cast steel segments fitted inside the old brickwork, and grouted up. When this had been done, it was possible to reduce the thickness of the old brick crown from 4 ft 6 in. to 1 ft 6 in., thus giving a depth of 31 ft through the cutting. The breaking out of the concrete and brickwork was slow and tedious. Shipping had to have a free passage, and the swing bridge taking road and dock rail traffic had to be kept operable. The area to be deepened, some 100 ft long, 80 ft wide and 3 ft thick, was to be broken out by divers. The old brickwork was very hard, and about 33% of the working shift was lost because of craft and shipping movements. The output achieved averaged one cubic yard per diver each week.

To speed up operations a special diving bell was built, 10 ft 6 in. long, 6 ft 6 in. wide and 6 ft high, equipped with electric light, telephone, an air-line, and high

Detail of lock gate, King George V Dock

pressure air for working compressed air tools. There were seats inside for the men to sit while being raised or lowered to the dock bottom. Later an overhead runway was fixed from which skips were hung to take the excavated material.

Using the diving bell, progress was greatly improved to 15½ cubic yards per week with four men working inside. Another benefit of using the bell was the frequent and easy inspection of the work it made possible: it was most important that no high spots were left which could later damage ships passing through the cutting to Victoria Dock.

COMPLETE REBUILDING OF VICTORIA DOCK

While the Connaught Road cutting was being deepened, work began on re-modelling the Victoria Dock. The five finger jetties, once the pride of Victorian engineers, were considered to be inadequate, too frail, too short, and no longer safe. They had been constructed with horizontal arch panels of 14 in. brickwork between cast iron T columns 7 ft apart. These were held in place by two tiers of 2½ in. dia. tie-rods. They could not be inspected, so their condition was only known, too late, when the walls bulged or collapsed.

The jetties were unable to deal with cargo vessels lying on each side, thus only half of the quay could be used at a time. The warehouses below the quay level had never been watertight, and were frequently damaged by ships berthing heavily alongside. After 80 years of continuous use it was felt that the time had come to demolish them, and build a straight quay with large modern warehouses behind as had been suggested in the 1910 plan.

As the development of the north side of Victoria Dock would mean that the dock would be closed to shipping due to reconstruction work, it was decided to moder-nize the south side of the dock first.

Although many of their sheds had been repaired after 1917 when ammunition being loaded at Silvertown exploded causing much damage, the old timber quays in the southwest side required rebuilding. This was 1750 ft long and was recon-structed using cylinders and a suspended quay in a similar manner to the North Albert scheme. A modern three-storey reinforced concrete framed transit and storage warehouse 504 ft long and 120 ft wide was built in 1935. This was followed by two similar sheds afterwards providing three first class berths.

The 27 acres at the southeast corner of the Royal Victoria dock had remained undeveloped owing to the dumping of a large quantity of excavated material to a height of 15 ft above the surrounding ground. Possibly this had arisen when the pontoon dock nearby had been constructed in 1855. In 1936 the first stage of the main quay construction was carried out using 60 concrete monoliths in the manner earlier adopted at Tilbury to deal with poor ground. This gave a main quay 1200 ft long, with a return quay to the pontoon dock 560 ft long.

Once the south side of the dock had been tidied-up the way was clearer to com-mence the major work of reconstructing the north side finger piers in 1937.

The scheme adopted consisted of a continuous open type reinforced concrete quay 3250 ft long sited so as to give sufficient room for a 150 ft wide warehouse, the railway tracks and the roadway. It involved the removal of the southern end of the jetties, and reclamation of the water area on which five modern three-storey ware-houses were to be erected.

The contract commenced in January 1937 and as in the case of the north quay to Albert Dock it was decided to work from both ends simultaneously.

Construction consisted of three rows of cylinders with strong transverse beams supporting 14 in. thick concrete sheet piling to retain the fill material behind forming the warehouse areas. A deep coping beam was constructed on a precast

29

concrete soffit shutter supported on the outer row of cylinders at impounded water level. This was considered to be an important factor in preventing the swim ends of loaded barges from getting trapped under the quay. When the transverse beams were in position it was impossible to get under the quay to strike conventional deck shuttering, so a system of precast concrete slabs 4 in. thick was laid, supported by the beams. A 12 in. thick slab could then be constructed in the normal way.

When this work had been completed, it was possible to dredge the dock to give a depth of 31 ft, matching the new depth of the Connaught Road cutting. Although the construction of the five warehouses could not be completed until after the 1939–45 war, the modernized Royal Docks remained almost unchanged until after their partial closure in 1978.

Other improvements, 1910–1939

DEVELOPMENTS AT TILBURY DOCK

Although the newly appointed Port of London Authority had concentrated on improving the dredged channel, and the Royal Docks, it was found necessary to carry out modernization to other docks as well.

In the last years of the 19th century when dredging of the Thames had been neglected, the Orient Line decided to move its base from Albert Dock to Tilbury. They maintained that a giant like their 5368-ton *Orient*, the largest ship then using London, could no longer be taken up the river in safety. It was the start of a long association between Tilbury Dock and the routes to Australia and the Far East.

The Pacific and Orient (P & O line) also required accommodation and it was decided to extend the main dock westwards by 1611 ft. This would provide three berths, each with a transit shed 600 ft long by 120 ft wide. These were to be provided with railway tracks at the rear, so that passengers could join the boat train at St Pancras Station and travel in comfort right up to the shed. They would then be cleared by customs to join the ship alongside without having to walk more than 100 yd.

In August 1912 a contract was let to Topham Jones and Railton for the construction of the dock extension. This was to have a quay wall on the south and west sides only, the north side being left as a chalk pitched earth bank ready for more extensions should they be required in later years.

When the original dock was constructed it was found possible to excavate in the dry almost to the sand and gravel 40 ft below quay level. Only in the lower part of the trench for the dock wall was it necessary to use timbering. In the 35 years between the construction of the original dock and the proposed extension, considerable softening of the clay must have taken place. This was probably due to the fact that the impounded water level in the dock was some 12 ft above the natural ground water table, and at the west end there was no dock wall, only a pitched slope.

The excavation 262 ft west of the old dock wall had reached a depth of 15 ft below Ordnance Datum (OD) when the contractors ran into difficulty as the ground began to flow into the trench from below. Steel sheet piles were adopted, but it was found that their lower ends buckled inwards and the amount dug out during the day was lost the next night. However, the contractors struggled down a further 7 ft to 22 ft below OD, but by then the inflow of ground was such that the piles bent inwards to a radius of 60 ft. At this stage work was abandoned, the piles were withdrawn and the excavation was backfilled with gravel.

The solution to the problem was found in changing to a 'monolith' form of construction. After the ground had been levelled out along the proposed line of the new dock wall, a weak concrete base was laid. On this a precast concrete shoe with a cutting edge was placed. The sections were made up of concrete blocks each weighing 5 tons. These blocks made a square with 30 ft sides, with four 10 ft square wells inside. The walls were 3 ft 4 in thick, with radial joints to resist earth pressure.

When the sections had been built up to some 12 ft high, the concrete base was cut away and they were allowed to sink into the ground under their own weight. Material inside the wells was removed by grab, and more precast sections were added until the skin friction on the outside of the walls prevented them sinking further. Kentledge blocks made of cast iron were piled up on top of the last section, and sinking continued until the toe was at the required depth in the gravel.

Verticality was controlled by grabbing out as necessary to correct any lean to one side or the other. The monoliths were pitched with a 6 ft wide space between them, so that each could be sunk independently of its neighbour. The space left between the monoliths was afterwards grabbed out and filled with concrete. After they had been sunk to their correct level, they were cleaned out, and the bottoms sealed with a concrete plug placed under water. The back two wells were then pumped out, and filled with concrete to increase the stability against the earth pressure behind. The sinking took 25 months from September 1913 until October 1915. A total of 2330 ft was sunk, at a rate of 21.57 ft/week. The whole contract was completed in August 1916.

The system of building dock walls in this manner was so successful that it was adopted for all further extensions at Tilbury. The dimensions of the monoliths have remained unaltered, only the methods of sinking being improved.

TILBURY RIVER CARGO JETTY

During the period that the P & O berths were being constructed, it was decided to embark on an entirely new venture. Some of the larger ships were in the habit of calling at Tilbury, unloading part of their cargo, then crossing over to the Continent to unload the remainder. If a cargo jetty were to be constructed in deep water out in the river, considerable time would be saved in locking into and out of the main dock.

The Tilbury river cargo jetty was 1000 ft long and 50 ft wide. It was equipped with storage warehouses below dock level, and with railway tracks on the deck, and goods could be unloaded and distributed quickly to the main line railway net-

First use of monolith construction (1914), Tilbury

works. The jetty was located so that barges could tie up on the inside for loading goods to be taken up river to the London and St Katharine Docks.

Work commenced on the construction in 1913 but it was not considered to be of sufficient priority to warrant labour being employed on it during the war and it was not until 1921 that it was opened to shipping.

Although at the planning stage it was considered to be a sound proposition, it was never popular with the shipping companies. At Tilbury there is over 20 ft rise and fall in the tides with a 4 knot current. The level of the ship's deck in relation to the jetty would change by over 4 ft in an hour. This would necessitate constant adjustment to the mooring lines by the ship's crew.

FLOATING LANDING STAGE

To overcome this difficulty it was considered that a floating landing stage would be of advantage, particularly where passengers were to be embarked or disembarked. The level of the ship would remain constant in relation to the landing stage and adjustments caused by the rise and fall of the tide would be taken up by long ramps from the shore.

The scheme was authorized to go ahead in 1922 and was welcomed by the shipping companies. Until the stage was opened by Prime Minister Ramsay MacDonald in June 1930, ships which did not land or embark their passengers inside the dock had to anchor in the river. Tugs or river steamers acted as ferries from the ship to small piers on the shore. Although this method may have added spice to the voyage for younger passengers, it was an inconvenient and time-consuming method of embarkation.

The new floating landing stage was 1140 ft long, and was capable of accommodating the world's largest liners. The downstream end was reserved for the Tilbury to Gravesend ferry and for excursion steamers. It was situated downstream of the dock entrance tidal basin and close to the existing Tilbury Hotel.

The landing stage was constructed of a number of large steel tanks held in position by link arms attached to massive anchorages in the river bank. The deck was at a constant height above the water, and covered movable gangways were provided to fit the doors of any liner coming alongside.

On the shore there was an imposing cathedral-like baggage hall for customs officials and immigration officers, intended to give the newcomer to Britain the

Cargo jetty and floating landing stage, Tilbury

impression of a country stable and wealthy. Only a few yards away a car park was provided for those passengers being met by road.

Tilbury Riverside Station provided a fast boat train service to St Pancras or Kings Cross Stations in the City. Hundreds of passengers could be handled speedily without confusion, a testimony to the careful planning of the whole landing stage lay-out.

NEW ENTRANCE LOCK AND DRY DOCK

In the 1920s deepening and widening of the Suez Canal was carried out. As Tilbury had become established as the home port for the Far East and Australian trade, it became necessary to provide additional facilities for the larger ships, which this made possible. In 1926 a new entrance lock costing £1 600 000 was commenced. It was situated between the west end of the main dock and the river. Although this required ships to turn to enter the lock, this manoeuvre had always been accomplished without difficulty. The lock, 1000 ft long and 110 ft wide, was the largest entrance in the Port of London, and was built using the monolith principle found so successful for the quay walls for the P & O berths.

At the same time a new dry dock of ample proportions was built to service the new larger ships. This was 752 ft long and 110 ft wide with a depth of 47 ft 6 in. from floor to coping. Both these contracts were carried out by Sir Robert McAlpine, who made an impact on the local inhabitants. Some older workers in dock maintenance recalled stories of the difficulties experienced in construction. It is difficult to separate rumour from fact, but there was possibly at least one occasion when a locomotive with a train of skips full of concrete became derailed and fell into the bottom some 50 ft below.

With the completion of these works and the Passenger Landing Stage, Tilbury Dock was well established as the terminal for the Australian traffic. The dock remained unaltered until further developments 25 years later.

THE TRINITY SQUARE HEAD OFFICE

The newly formed Port of London Authority was most anxious to gather together under one roof all the staff scattered around the various docks in small cramped offices. This was seen as an essential factor in welding together the differing traditions of the old Dock Companies into a uniform and efficient administrative organization.

The choice of a site was important as it would need to be near to the city and the head offices of Shipping Companies. For easy access to West India, Millwall, the Royal Docks and Tilbury, it would also be of considerable advantage to be near Fenchurch Street Station.

One of the sites considered was that of the Crutched Friars warehouses, originally built by the East India Dock Company and taken over as part of the London and India Dock Company's estates at the time of transfer. The old warehouse was situated on Tower Hill, near the heart of the City and Fenchurch Street Railway

Station. It was an historic site, having once been occupied by the old Navy Office from which Samuel Pepys had watched the Great Fire of London in September 1666. His own house had been in Seething Lane nearby. Trinity House was on the east side of Savage Gardens, and opposite was the site of the Tower Hill scaffold upon which many State offenders had been executed. The area is now part of Tower Hill Gardens.

The warehouse occupied 1½ acres and it was necessary to acquire the buildings in the area to make an ideal site some 3 acres in extent. The important frontage to Trinity Square could be made into an imposing architectural feature. To obtain the best design it was decided to hold a nationwide competition open to all architects. Sir Aston Webb, President of the Royal Academy, was appointed to advise and act as assessor; 170 different designs were received from which Sir Edwin Cooper's layout was finally unanimously selected.

The Trinity Square Office was built in the form of a hollow square having one corner cut off by the main entrance with its Corinthian pillars and sculptured figures. The central area used as the main public office and known as the Rotunda was roofed by a magnificent dome. Unfortunately this was irreparably damaged when a bomb fell into the centre lightwell in November 1940.

Construction of New Entrance Lock, Tilbury

PLA

New Dry Dock, Tilbury

Sir Edwin Cooper, like other eminent architects of the time, not only entrusted the work to his selected builder, but insisted on having his own choice of general foreman. John Mowlem and Company Limited, who had a close association with him, and who had established over the years a high reputation for good workmanship, were the Contractors.

A considerable amount of demolition and site clearance had to be carried out before the main building could begin. The outbreak of World War I delayed construction, but the foundation stone was laid by Lord Devonport on 23 June 1915. Steel was in short supply for the massive steel framework (which later resisted enemy bombing so well in 1940) but finally seven years later it was completed and opened by Prime Minister Lloyd George on 17 October 1922.

RAIL COMMUNICATIONS

The railway networks of the 19th century played an important part in the communication systems developed between the shipping companies' headquarters in the City and their Ship's Masters in the docks.

London and St Katharine Docks were close at hand, and messengers on foot could reach their destination in a comparatively short time, but West India Dock was further away. In the days before the telephone had been invented, it was necessary to have a quick means of getting to the dock, and in April 1849 the London and Blackwall Railway began a service between Fenchurch Street Station and West India Dock, with trains every 15 minutes on week days, and at half-hourly intervals on Sunday. Trains ran until the General Strike of May 1926.

The line was extended into the Isle of Dogs to the Millwall Junction and North Greenwich when the new Millwall Dock was built. The arrangements were most complicated. The track as far as Poplar Dock was owned by the Great Eastern Railway. The next half mile was owned by the East and West India Dock Company, later amalgamated into the London and India Dock Company, with a station near South Dock entrance. The crossing of the entrance lock was by a narrow swing bridge which limited the size of railway carriages and meant that until 1880 only horses could be used to draw them, though light 2–4–0 tank engines were later used. South of the station the track was owned by the Millwall Dock Company. Here it ran on to a viaduct on which Millwall Dock Station was built. Beyond to the terminus at North Greenwich it was owned by the Great Eastern Railway. In spite of these complicated arrangements trains ran regularly without confusion. Dockers and shipping clerks kept the line busy, and in 1922 the PLA bought some Great Western Railway long wheel base railcars to improve the service still further. The railcars comprised two coaches permanently coupled together, with a vertical 'coffee pot' boiler in one connected to steam cylinders driving the front set of wheels. Unfortunately most of the track has vanished without trace. Only a portion of the viaduct remains southeast of the transporter yard to remind the local inhabitants of the days when the Isle of Dogs was provided with a fast and regular rail service to London.

The Royal Group of docks was always supplied with good rail services. Thomas Brassey had built the North Woolwich line specially to serve his new Victoria Dock, and the sole remaining railway link with North Woolwich was most popular when river trips became fashionable in the 1880s and 90s. Today with its station at Silvertown it still serves the refineries of Tate & Lyle as well as the south side of the Victoria and Albert Docks.

With the building of Albert Dock and the Beckton Gas Works, special branch lines ran from Liverpool Street and Fenchurch Street via Stratford and Canning Town to Gallions Hotel and Beckton Gas Works, with stations at Tidal Basin, Custom House, Connaught Road, Albert Dock Central, Manor Way and Gallions on the north side of the Docks. The branch line to the gas works was built by the Gas Light and Coke Company, the branch line to Gallions by the dock company, and the Great Eastern Railway ran two trains an hour from Fenchurch Street. Passenger trains ran until 8 September 1940 when the line was closed by the heavy air raids on the docks. After World War II it was reopened for goods traffic, but finally abandoned on 17 April 1966, when no further goods were accepted by rail in the docks.

Tilbury Dock was built with the railways in mind. There was rail access to each quay, and special boat trains ran to the P & O berths. It was also served by Tilbury Town station, and the Riverside station on the London and Tilbury and Southend line.

Only Surrey Docks were without rail access. The nearest station was Surrey Dock Station on the New Cross-Whitechapel branch of the Underground Railway which crosses the river in Brunel's famous Thames Tunnel. As most of the trade was timber, transported by barges to wharves along the Surrey Canal, or on the Thames, railway connections were not so essential to the smooth running of these docks.

MORE DEVELOPMENTS IN LONDON AND ST KATHARINE DOCKS
The whole of the land space in St Katharine Dock had been covered by tall warehouses, when built by Thomas Telford, leaving no room for improvements by the

Mowlem

PLA Trinity Square Head Office

PLA. In London Dock, however, it was possible to increase the capacity of the western dock, at the expense of the loss of some water area.

It was decided to build a jetty out from the west wall of the dock to berth two ships each side, and one across the end. A two-storey transit shed and warehouse was built on the jetty. In addition it was decided to build a false quay in front of the old dock wall with similar facilities for transitting and warehousing.

The design in 1911 was in the newest material-reinforced concrete using the Hennebique system. Construction was carried out by the firm of L. G. Mouchel, who at that time were pioneers in this form of construction. The construction was still in perfect order when the docks closed some 56 years later.

The vast warehousing complex of Cutler Street warehouse, St Katharine, and London Docks was to be developed as a world centre for valuable merchandise. Almost all the goods stored in the warehouses of St Katharine Dock were transferred from the lower docks by lighters. The largest warehouses E and F (both completely destroyed by enemy bombing in 1940) had, in addition to the ground floor, seven upper floors and vaults underneath. Colonial wool and mohair were stored there, forming part of the London and St Katharine Docks wool department. Covered-in bridges joined the warehouses across Thomas More Street to the wool warehouses in London Dock. F vault for many years was the centre of the London rubber trade.

Included in a long list of other goods stored were: dried fruits, shells, hops, rags, marble, sugar, indigo, tallow, glucose, matches and tortoiseshell. The shed on the south side of the dock contained a tank for live turtles.

Nearly two-thirds of the nation's tea was housed in the warehouses near to Tower Bridge. In one occupied by a well known tea merchant, as late as the 1950s, tea was blended by tipping the contents of a number of tea chests out onto the floor, and turning over the pile by hand shovel before resealing it into chests with fresh labels.

Cutler Street Warehouse, opposite Liverpool Street Station, once the fortress of the East India Company, had 15 acres of floor space. It has been described as a 'priceless treasure house'. Here wines, spirits, tobacco, opium, and carpets were stored.

After the collapse of the Turkish Empire in 1918, Cutler Street held the unique position of being the market place for carpet distribution, where carpets from all over the world were held for re-export. Tea was also stored here, a trade which grew rapidly with the increasing popularity of Indian and Ceylon blends.

London Dock between the wars had 2¼ miles of quays, 22 acres of wine vaults and in the warehouse 3½ million sq. ft of storage area. Long term maturing of brandy took place in the brandy vaults. Vintage port was carefully bottled by hand in vaults illuminated by candles. In all, wines and spirits valued at over £18 million were kept in the vaults.

Other goods stored were wool, skins, ivory, spice, rubber, mercury, essential oils, fish oils, gums and glues, drugs, canned fruit and fish, dried fruit and nuts, sugar, coffee, cocoa, paper bark, hemp, coir yarn, jute, seeds, coconuts, whalebone, canes, tallow, mats, wax and iodine. Up to 200 tons of ivory were sold each year. It was the main warehouse of the City of London, and with the ample room for display, goods were not only stored but shown so that the best prices could be obtained for the merchandise offered. The PLA officers were regarded as some of the world's leading experts. The spice floor had experts able to ensure that the quality of goods offered for sale were the best obtainable.

In the 1920s and 1930s there would be up to 30 ships in the dock at any one time with a gross tonnage of about 25 000 tons. It would have been considered a viable small port on its own.

The increase in size of ships, the loss of a considerable amount of warehousing during the War and the increase in specialization caused the dock to lose money, and the later change to large container ships saw the end of the vast warehousing complex by 1967. Only the carpet trade remained, rehoused in No. 10 warehouse London Dock.

WEST INDIA AND MILLWALL DOCKS

West India, the longest established enclosed dock, received little immediate attention from the newly formed PLA who preferred to concentrate on the modernization of the Royal Docks. However, in 1911 Mouchels modernized the north quay of the Import Dock in a manner similar to the north quay of London Dock. A false quay was built in front of the old wall over the water, and a two-storey transit shed and warehouse built behind. This quay was always operated by the PLA themselves with permanent staff.

In Millwall Dock, the Central Granary had been built by the Dock Company just before the formation of the PLA. Mechanical grain handling had been developed in 1893 and a good trade had been built up with Black Sea and Russian ports. This Asian cereal was softer than that grown in America, and the experts preferred open floors with good ventilation for this type of grain. As a cargo might comprise a number of small parcels, it was necessary to keep them separate.

At the time there had been a number of disasters with vertical grain silos, particu-

Exchange sidings, Victoria Dock

PLA

PLA

Mowat's Central Granary, Millwall Dock

larly due to the falling grain arching and jamming, then suddenly dropping with considerable force to the bottom. Magnus Mowat, a civil engineer, thought of a new method which combined the advantages of mechanical handling with the requirements of horizontal storage.

Three pneumatic grain elevators were erected on a dolphin 350 ft long and 25 ft wide, some 50 ft from the quay. These could discharge directly into barges moored inside the jetty or to the granary on the quayside. The granary, designed to hold 24 000 tons of bulk grain, was 250 ft long and 100 ft wide. It was brick built, with 11 floors for storage and inspection, and a delivery floor and basement below. It was divided into five compartments with vertical fire walls, and a 20 000 gallon tank on the roof provided water for instant fire fighting. Windows gave good ventilation to limit fire and explosion risks. The grain was distributed to the various floors by elevators and horizontal conveyors, and stored to a height of 4 ft 6 in. Bulk boards were provided to keep the samples of grain separate. It was built for sacked delivery to horse-drawn carts, but its capacity and flexibility was such that it was possible to deliver to large bulk carriers in later years without effort. The cost was £170 000 and as far as is known, no other granary had been built like it.

It was not until 1927 that it was decided to rebuild the West India entrance lock which was by then over 100 years old. In August of that year a contract valued at £647 600 was let to Sir Robert McAlpine. Unlike Tilbury the ground was good, and using the recently developed steel sheet piling system the walls for the new entrance lock were built in trenches without difficulty. While this work was in progress, it was decided to widen the bellmouth to the Import Dock, and to join Millwall Dock to West India by a new cutting. This work was let to Charles Brand for £390 000.

With the completion of these schemes in 1928, access to the Docks was greatly improved. For the first time shipping could enter Millwall Dock from the east side of the Isle of Dogs rather than spending time in going further upstream to the more difficult Millwall Entrance. It was fortuitous that this work was carried out, as in 1940 the Millwall lock suffered a direct hit by enemy action, and was permanently closed.

In 1935 a disastrous fire completely destroyed the old Rum Quay warehouses. When the debris had finally been cleared away, room was made for a redevelopment of the south side of the Import Dock. A start was made by widening the quay for the first berth. This was done by sinking cylinders into the dock bottom in front of the old quay wall in a manner similar to that employed in the remodelling of Victoria Dock. A new two-storey transit floor with a warehouse above it was built at

Reconstruction of Rum Quay, West India Dock

PLA

Canary Wharf for the fruit trade. This work was completed by 1939 but further developments had to wait until 1950.

SURREY COMMERCIAL DOCK

As mentioned in the previous chapter, Surrey Commercial Dock had remained aloof from the intense competition between the 'North Bank' docks during the 19th century. It was left unchanged by the Port of London Authority until 1925, when Sir Edwin Cooper put forward designs for new timber storage sheds. These were steel frames with galvanized corrugated steel roofs. Large overhanging canopies at the gable ends protected them from the weather but the sides and ends remained open. Each timber 'deal' was separated from the next in the stock by a small batten spacer so that a current of air could circulate to ensure perfect drying and seasoning.

In order to minimize fire risk, massive brick fire walls were incorporated in the construction, dividing one section from the next. New timber storage sheds were built in Acorn and Albion yards first, whilst the old Grand Surrey Timber Dock was reshaped and deepened to form Quebec Dock. When this had been done more timber storage sheds were built in Centre Yard. In all 33 000 standards (1 Petersburg standard is 165 cu. ft or about 2½ tons of softwood) could be stored under cover, and a further 42 000 in the open.

And so it was that Surrey Dock became recognized as one of Britain's major timber trade centres.

With the development of packaged timber requiring special mechanical handling and large storage areas behind each berth for speedy unloading, the work of the timber porter vanished. The Surrey Dock was no longer suitable, and the trade moved to the new branch dock at Tilbury.

It was felt by many of those who had spent their lives in the timber trade that this was a retrograde step. They liked to be able to visit the timber ponds (some softwoods benefitted by storage under water) and select their own logs. They also enjoyed wandering through the acres of timber sheds, seeing the joinery timber being properly seasoned under the watchful eye of the PLA staff.

OUTBREAK OF WAR, 1939

By 1939 the modernization schemes carried out by the vigorous Port of London Authority had welded together the old dock companies into an efficient and well co-ordinated port. Trade in the 30 years of the Authority's control had increased by 50% to 60 million tons annually, including private riverside berths. This amounted to 38% of the UK total, and was the largest quantity handled by any port in the world.

The introduction of mechanical handling had led to the number of men employed being reduced in the same period by 18 000 from 52 000 to 34 000 but the outbreak of war on 3 September 1939 put an end to any plans for increasing the capacity of the port still further.

44

World War II, 1939–1945

MAGNETIC MINES

At the start of the Second World War London was the premier port in the United Kingdom, but within two months Hitler's first 'secret weapon' had been disclosed. Large quantities of magnetic mines were laid in the estuary, sinking some ships and causing others to be diverted to west coast ports. At this time, before the fall of France and with the Navy in command of the Channel, no thought had been given to protecting London.

However, the magnetic mine menace was soon overcome. By wrapping large cables around a ship, and passing electric currents through them, the ship's magnetic field could effectively be cancelled. For this, 'degaussing' stations, including one at Tilbury, were set up and by early 1940 London was back to normal after major minesweeping had been carried out.

A boom was constructed across the river between Minster in the Isle of Sheppey and Southend, to prevent enemy submarines penetrating the upper reaches of the river. To add to the defence of the estuary six concrete forts were constructed in Surrey Dock, carried on dumb barges and lowered by the falling tide on to prepared piled foundations. They were sited on the Shivering and Red Sands near Whitstable, and were equipped with anti-aircraft guns.

DUNKIRK

At the time of the evacuation of the British Expeditionary Force from Dunkirk, the London docks and river played an important role. From 20 May 1940 until the evening of 26 May when operation 'Dynamo' was put into action, the Harbourmasters and Dockmasters helped organize the Armada of small craft and the first troops brought home. As well as privately owned craft being gathered, every boatyard from Teddington to Brightlingsea was searched, and more than 40 serviceable motor boats and launches were found. Lifeboats were taken from ocean liners in the Royal docks, and Thames barges pulled by tugs were also pressed into service. By the night of the 27 May over 400 small craft were at sea, many of them ferrying men from the beaches at Dunkirk to larger ships anchored in deep water. Nearly

100 000 men were taken off in this manner with the loss of 170 small craft.

When he became Prime Minister Winston Churchill had foreseen the vulnerability of London to enemy bombing, but even after Dunkirk and while the Battle of Britain was being fought over Kent and Sussex little attempt had been made to disperse the vast stores of food and timber in the dock and riverside warehouses. Up to this time a quarter of all England's imports still came through the Port of London. A fifth came through Liverpool but only one-tenth each through Southampton, Bristol and the Humber ports.

AIR RAIDS

The bombing of London began with 68 aircraft on 6 September 1940, in broad daylight, followed by the first large scale attack by about 300 planes on 7 September. London docks were an easy target for bombers in a daylight raid. They covered such a large area that any bomb, however indiscriminately released, was bound to find a target.

Saturday 6 September 1940 was a quiet sunny day. It turned out also a lucky day for the men working at Basin South Depot, Royal Albert Dock. They were expecting to work until 4 pm but because of an uneasy feeling, the foreman allowed them to go home at mid-day. The depot at Basin South was the target for one of the first bombs in the air raid. Soon the carpenter's shop, timber stores and offices were

Mowlem

Wartime forts at Shivering Sands

46

PLA Collection, Museum of London

Raid of 7 September 1940

completely burnt to the ground, all records being lost. The air raid shelter where the men would have taken cover was demolished by a direct hit from an aerial torpedo.

In Surrey Commercial Docks large quantities of timber were stored for seasoning in open-sided sheds, each piece being separated from the next by small battens to allow air to circulate and so dry the timber. After the hot summer of 1940 it was in an ideal condition to catch fire when bombing started.

The fire of 6–7 September was so intense that the glow could be seen in the night sky as far away as Guildford. It was the fiercest ever recorded in Britain and when the night bombers followed up they needed no sophisticated ray device to guide them to the target. The timber sheds in Russia and Quebec Yards and Surrey Dock were completely burnt out, as also were those in Millwall Dock. Only the heavy bulk timbers in the ponds of Lady and Lavender Docks were saved.

In London Dock large quantities of rum stored in barrels caught fire and the intense heat of the burning spirit set stores of paint and rubber alight. The pall of black smoke made fire-fighting impossible. St Katharine Dock House was demolished. On the river, barges caught fire and drifted out of control. In places the surface of the river was a sheet of flames where burning molten sugar had flowed over it.

From 7 September until 3 November about 200 bombers attacked London each night. The worst time of the early blitz was 15 October when 480 bombers dropped 386 tons of high explosives and 70 000 incendiary bombs. During this period much of the tea kept in St Katharine Dock warehouses was destroyed. Stores of butter and flour were also lost. In the four months between September and the end of 1940, 13 339 people were killed in London and a further 17 937 injured. In all about 30% of all warehousing in the London and St Katharine Docks was completely destroyed.

After 3 November the German bombers transferred their activities to Coventry and other manufacturing cities. This gave London a welcome break, and time to clear away bomb rubble and repair sewers, water mains and other services which had been put out of action. As St Katharine Dock was so badly damaged, it was decided to use the inner basin as a dump for bomb rubble from the city and buildings nearby.

During this time it was decided to fell the two Victorian chimneys of Abbey Mills pumping station at Stratford. They had become redundant with the modernization and electrification of the pumps and were considered to be outstanding landmarks to guide German bombers to their targets in the dock area. However, their removal from the skyline appeared to have no effect on the accuracy of further raids.

Fires at Surrey Docks

PLA Collection, Museum of London

48

Derby Aerosurveys

St Katharine Dock, after war damage, Dock House, A, D, E and F warehouses completely destroyed

Emergency fire mains had been installed in the city with an elaborate pumping station at Dowgate Dock close by Cannon Street Railway Station. The pumps drew raw water from the River Thames through an intake and tunnel under the river bed which delivered water at low pressure into a 3 ft dia. main.

Unfortunately the firemen had not been adequately instructed in the use of the emergency mains. On 29 December after a quiet Christmas holiday the bombers returned to London with a classic incendiary raid on the docks and city areas. Heavy parachute mines were followed by a massive incendiary attack which started 1500 fires in the city and upper dock areas. Firemen connecting their hoses to the new hydrants found that the pressure was low and decided that the mains had been broken.

Chaplain's wine warehouse on Tower Hill was badly damaged. A six-inch diameter pipe ran continuously into the river with wine from smashed casks for four

49

days. The sight was galling at a time when wine was practically unobtainable by the general public.

The Guinness warehouse in the Eastern Dock, London Dock was full of spirit which went up in flames, producing a fire so intense that the cast iron columns actually melted. When the warehouse was rebuilt after World War II one was preserved.

EVACUATION OF LIMEHOUSE

Limehouse, the heart of Chinatown, was situated near to West India Dock Entrance. Character was given to it by the row of old 'Pucka-Poo' houses or opium dens to which Thomas Cook used to run tours before the war. Chinamen in national costume and pigtails would run out of the houses chased by another brandishing a large meat chopper. The tourists used to stand open-mouthed at the show pre-arranged for a fee of 2/6d [12½p] per performance, paid in advance by the coach driver. Early in November 1940 the whole area was severely damaged and Chinatown has never been the same again. The uninjured Chinese were collected to be rehoused in Liverpool. One old man missed the coach, and was found later uninjured in the wreckage of his hovel, still in a happy opium stupor and oblivious of the bombing going on around him.

Mowlem

Orange Shed, Victoria Dock, column failure caused by bomb blast

The heavy bombing made London docks unusable for further ocean shipping. Three-quarters of its trade and many of the dockers moved to Liverpool and to the Clyde where an emergency port had been constructed.

After the Spring of 1941 bombing became less frequent and it was possible to clear up a lot of debris. Vacant berths and dry docks were taken over by the Navy for the repair and overhaul of all types of warships.

PETROL BARGES

In 1943 the docks were put to use for additional war effort. The plan was to take the war into Europe, producing all the problems of landing an army on the Normandy coast and of keeping a mobile force supplied with fuel. The first solution was to provide petrol carrying barges on similar lines to enclosed Thames barges. These could be towed across the Channel in 'strings' to the landing areas. The shortage of both steel and skilled barge builders was a major problem and concrete was tried instead. High quality watertight panels were made under factory conditions by Messrs Wates the London builders, and an ideal site was found at the old Rum Quay, West India Dock for building them.

The panels were assembled and held in jigs whilst being concreted together. When set the 'London Mammoth' (the PLA 150-ton capacity heavy lift crane) was called in to put them in the water. When suspended about six feet in the air they were inspected underneath for watertightness before being lowered into the water. The barges were superseded by PLUTO (Pipe Line Under The Ocean, see next section, page 61), but they were moored as reserves near Lambeth bridge for a number of years after the war.

THE MULBERRY HARBOUR AND PHOENIX BREAKWATER

Others have written about the floating harbour that went to France at the time of the D-day landings under the code name 'Mulberry'. Little if anything has been said about the difficult conditions under which the 'Phoenix' units forming the breakwaters were built and the part played by the Port of London in providing facilities for construction and advice on towing the units.

By the end of July 1943 the 'Overlord' plan for the invasion of France had been accepted by the Americans and British at the Quebec Conference. There would be a joint landing on the Normandy coast, and to make it successful, two harbours would be established – 'Mulberry A' for the Americans, and 'Mulberry B' for the British. Floating breakwaters would be towed across the Channel in sections and sunk on to the sea bed, so that ships could unload in sheltered waters at all times however rough the sea outside. Each harbour would be about the size of that at Dover.

The scale of the scheme was immense. Time was unbelievably short, about six months in which to plan, design and construct 147 units requiring over half a million cubic yards of high quality concrete, and this after four years of wartime conditions with an acute shortage of labour and materials.

To attain this most important objective Winston Churchill appointed Sir John Gibson to the position of Deputy Director General of Civil Engineering Works (Special). He was the driving force behind the civil engineering contractors, George Pauling & Company Limited, and had a reputation of being a 'go-getter'. He had as his assistants two civil engineers as Directors and Co-ordinators and Chasers.

Early in September a meeting was called to set up a joint committee of Contractors and Consulting Engineers to consider an outline design for the breakwater. Sir Malcolm McAlpine chaired the Contractors Committee and drew up plans for something that could be constructed within the short time allowed. The Consultants Committee turned the sketches into an outline design: the detailed drawings were to be prepared by the War Office. A list of basic principles was quickly hammered out as follows:

(a) the design had to be suitable for construction in the docks available, bearing in mind the limiting width and depth of entrance locks;

(b) it had to be simple and repetitive with no complicated beams;

(c) its requirements for labour and materials had to be within the capacity of the Ministries of Supply and Labour;

(d) the shape and size had to be within the towing capacity of tugs available, at a speed of 4½ knots;

(e) they had to be watertight and strong enough to stand hundreds of miles of towing in rough seas;

(f) they needed accommodation for a crew, and sufficient deck space for seamanship;

(g) they had to be capable of being sunk evenly and quickly;

(h) they had to be strong enough to sit on an uneven sea bed;

(i) they had to be strong enough to stand up to waves 120 ft long by 8 ft high without filling; and

(j) six miles length of breakwater needed to be built ready for towing to France, in a period not exceeding six months.

As many of the labourers would not be used to working at a height, it was felt necessary to have a deck at half-height to give a sense of security. The site organization, skilled supervision, and control of the works was to be the responsibility of the Contractors. Selected firms of Consulting Engineers would provide control, updating of the programme and co-ordination between Contractors sharing the same sites. The PLA Chief Designer was responsible for producing the working drawings in a period of under eight weeks.

The works were given the highest priority in the land, PML (Prime Minister's Letters), and the Ministry of Supply promised to make all necessary materials available, as required. Ernest Bevin, the wartime Minister of Labour, promised to scrape together about 22 000 men, although to achieve this 1000 carpenters were temporarily transferred from the Admiralty and a further 1000 from the Army.

Having decided that the landings would take place on the Normandy coast, possibly at Arromanches, it was necessary to survey the sea bed. This would provide information as to the depth of water, and to ensure that the proposed line of the breakwater was free from the protruding rocks which could have broken the backs of the units. This survey work was carried out in the utmost secrecy, mainly at night close to the German defences, the information being transmitted to the hastily assembled design team.

The results of the survey showed that the units in the deepest water would have to be 60 ft high, with others graduating down to 25 ft high near the shore. To aid towage the shape chosen was similar to Noah's Ark, with swim ends. A 6 ft wide deck would be provided just above the water line to enable a crew to handle towing and mooring lines when at sea.

The length of each unit was 204 ft. The beam varied between 56 ft 3 in. for the largest A units and 27 ft 9 in. for the smallest D units.

The A units weighed 6044 tons, and had an estimated draught of 20 ft 3 in. The intermediate 'B' units weighed 3275 tons and had an anticipated draught of 14 ft 6 in. The smallest only weighed 1672 tons requiring 13 ft 0 in. of water to float.

Finding suitable construction sites was the main problem. Ideally dry docks would have been the best solution.

Removable dam, East India Dock

For instance, the King George V graving dock at Southampton would have provided accommodation for nine A units. The dock could easily be dewatered, and the units floated out. However it had been commandeered for the construction of 'Bombardon', the floating booms envisaged as a first line of defence against storms whilst the breakwater units were being positioned. The dry docks at Portsmouth and at the Royal and Millwall docks, London, were full with the repair of cruisers, corvettes and destroyers and the Admiralty was not able to help.

The only dry dock available was the new one at Tilbury which would accommodate nine of the A units – not much use when about 147 units were to be built.

London had adequate accommodation, but the distance that the units would have to be towed was considerable and from London the Dover Straits would have to be navigated in the sight of enemy guns before the parking grounds at Dungeness could be reached.

As so often happens a compromise was reached. Seventy-six units were built in London and the remainder in the Southampton area.

To provide space for construction in London it was decided to dam the East India import dock, pump it out and build ten A units there. South Dock in Surrey Commercial Docks could also be dammed and dewatered, allowing a further eight B units to be constructed. These were the only small docks in which the units could be built full height in one operation. The remaining 49 units would have to be built half-height in basins excavated in the river bank alongside the Thames, at Erith, Barking Creek, Cold Harbour Point, Tilbury and Grays. Accommodation could also be found in the burnt-out timber yards at Quebec and Russia Yards, Surrey Dock.

DRY DOCKS AND BASINS

The units had to be floated out on the top of the tide and towed up to West India Dock for completion afloat in the deep impounded water there. Delay would be caused by the change in site, but it was considered to be the only solution to a difficult problem.

Sir John Gibson next took the unprecedented step of calling together the heads of 25 Civil Engineering Contractors. At this meeting he allocated the number of units to be built by each company and their site. The smaller Contractors were given up to four units each, larger firms between six and nine. The two most experienced London firms, John Mowlem and Sir Robert McAlpine, were allocated twelve and ten, respectively, and the Surrey and East India Docks for construction sites. The situation was unique. Contractors were given the opportunity not only to show the country what they could do, but how much better they were than their competitors working nearby.

The two contractors with wet docks had the most difficult problems. McAlpine in the Import dock at East India had the problem of building a concrete dam in the entrance capable of withstanding the water and any bombs which might drop nearby, and also capable of being demolished quickly once the units had been com-

PLA

Collapse of west wall (left) and south wall (right), East India Dock

pleted. Another problem was whether the old dock walls would be stable once the water had been removed from the basin. They were 120 years old and curved like bananas. In spite of the installation of ground water lowering equipment to relieve the water pressure behind them, the west wall collapsed three days after dewatering and on 18 January 1944 a further 300 ft of the south wall fell in. Nevertheless, these upsets were not allowed to hinder progress. Some of the units were repositioned to clear the demolished walls. A 6 ft thick layer of bomb rubble was laid in the dock bottom and well rolled to form a thick and firm base through which water could percolate later to provide the uplift needed to float out the units.

Mowlem were given two sites in the Surrey Commercial Dock. Eight units were to be built in the South Dock, and four in a basin to be excavated in Russia Yard. The South Dock was considered to be the best of all the Phoenix sites. The quay was served by hydraulically-operated cranes so that the walls of the Phoenix units could be built in large sections, the shuttering being lifted by crane, and there were modern warehouses for all ancillary work.

The snags in this otherwise ideal site were that the South Dock entrance had been

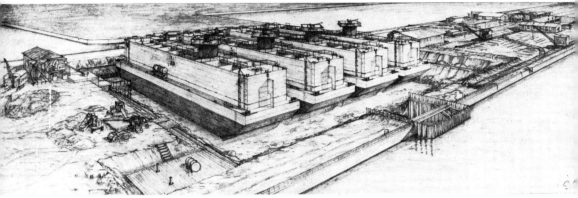

Phoenix units in Russia Yard Basin

badly bombed earlier in the war and had to be permanently dammed. The only exit would have to be through the Greenland Cutting which gave only 9½ in. of clearance – a tricky towing operation. Here also a dam would have to be built, but one capable of removal. As in East India Dock a layer of bomb rubble was used upon which to build the units, and over-enthusiastic tipping of the rubble at the west end of the dock on one occasion set up a mud wave which overwhelmed the dewatering pumps.

The team nominated to supervise the works had been building petrol barges in West India Dock, and immediately they had instructions to start, rushed to the firm's depot and claimed all the best plant available. The hydraulic cranes were slow in operation and when concreting was being carried out no steel fixing or shuttering could be done, causing major frustration and delays.

RUSSIA YARD

In Russia Yard basin in which Mowlem's were to build the other four units, the situation was very different. The supervising staff had been in East Anglia building an airfield where they had escaped air-raids. Returning to war-torn London on a cold and foggy October day to find the desolated area allocated to construction was an experience never to be forgotten.

The land chosen had the Lavender timber pond to the east and Russia dock to the west. The burnt-out remains of the timber sheds had been demolished down to ground level, but the foundations remained. A borehole was being put down to establish the nature of the ground, while various lay-outs were examined. The question was, would it be best to build two small basins with two units each, or one large basin in which all four units could be built together, though this would mean that the first to be completed would have to wait for the last to be finished before floating out.

The original intention was to build to half height in a shallow basin, flood, float out, and complete in the Russia dock alongside. However, the borehole revealed a thick layer of peat just at the level required for construction. This could have led to differential settlement during building, and damage to the units while the concrete was setting. Below the peat there was a layer of dense sand with gravel below, both waterbearing.

The contractor had to make one of the most difficult choices of the whole Phoenix scheme:

(a) abandon the site and look elsewhere (too late as excavation had already started);
(b) dig out the peat, and replace with bomb rubble (too time-consuming); or
(c) excavate the basin to full depth and install ground water lowering to enable construction to be carried out in the dry.

The last scheme appeared to be the most attractive as time could probably be saved by completing construction in one operation before floating out, particularly as flotation would be ensured by stopping the pumps and allowing the ground water to rise under the completed units. The sand layer would also make a firm base upon which to build.

Another problem was to decide how much clearance would be required for flotation under the units. Was the predicted draught, calculated as 14 ft 6 in., correct? No one engaged on the scheme had ever had to deal with large concrete floating craft but it was decided to allow 1 ft 0 in. During construction the anticipated draught kept going up and up as the design changed until it was thought that

Mowlem

Phoenix units afloat in Russia Dock

would be only 8 in. clearance left. However, the units finally floated quite happily with 10 in. clearance.

The excavated material had to be carted from Russia Yard out into Redriffe Road, over a swing bridge and finally tipped into the disused Quebec dock. Time was so important that up to 70 lorries were used.

The basin had sloping sides and everything had to be manhandled from the top of the bank. Concrete was supplied by four 6 in. concrete pumps mounted in pairs, each pair under a concrete mixing plant.

Men were drafted from as far away as Cornwall, and accommodation had to be found for them in addition to the usual canteen and cooking facilities.

As soon as construction had started, the German night bombers returned to London in the 'little blitz' of January 1944. The attacks lasted for three months, and although less severe than those of 1940, caused unrest among the labour force who wanted to get home before darkness fell and the raids commenced.

One night the General Foreman and a few of his trusted followers were finishing the day's concreting when all the lights failed, and the plant stopped working. The concrete pumps and pipelines were full of wet concrete which would set if the current was not switched on soon. The fault was traced to the main switchboard, where a bomb dropping nearby had caused a surge in the power supply and blown out the circuit breaker. With the aid of a long-handled broom the breaker was carefully replaced, and the current came back to normal.

Most of the civil engineering labour was of poor quality, those rejected by the Armed Services. They had travelled around the country for the last three years, living in labour camps away from their families and homes. Their work was either 'secret' or unglamorous. They wore no uniform, not even the green overalls supplied to agricultural workers. Many were unsuitably dressed and were given insufficient clothing coupons to buy stout clothes.

In spite of this, with the considerable driving force of the General Foreman, the construction of the first unit started on 5 January 1944, and was completed nine weeks later on 8 March. The units were all floated out on 17 March ahead of nearly all the others, and weeks ahead of those built in the perfect basin of South Dock. This achievement was celebrated by a 'floating out' party for the whole labour force.

By the end of March 83 units or nearly 60% had been completed. It was hoped that the remainder would be finished by the end of April, but it was early May before 100 could be sent to their parking places at Dungeness and Selsey. Both harbours were ready for the invasion by 23 May in spite of numerous design changes to the steel reinforcement, and delays in the supply of 3500 sluice valves and fittings.

There was a desperate shortage of tugs and towing gear for the units being taken from London, chiefly because many of the units were completed at the same time. Each unit was provided with an anti-aircraft gun, and quarters for the crew. One is reported to have shot down a German plane into the sea off Dover, when it tried to find out more about these curious objects!

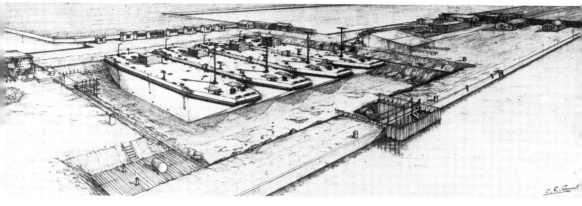

C. R. Smith

Admiralty barges, Russia Yard

The achievements of the Russia Yard team did not go unnoticed, as they were kept employed until the end of the war in Europe, and afterwards were transferred to rebuilding London Dock when peace returned.

Even before D day it was decided that it was probable that more units would be required. There were no reserves as the time schedule had been too short, and delays had prevented additional units being built before the landings. Early in May the Russia Yard team was sent to one of the by then vacated Phoenix basins to build a 60 ft high A unit half height to be taken to West India Dock for completion. The whole operation was completed in ten weeks, the unit being used at Arromanches to repair the breakwater when it was damaged by the heavy storms of 14–17 June.

Towards the end of July the temporary dam was reinstated at the Russia Yard site, and the basin pumped out to build three more B units, 40 ft high, similar to the first batch. These were required for what was called the 'Winterization' programme so that Mulberry could be kept operational in rough seas. The three units were built in nine weeks, but this time when the basin was flooded there was no sign of the units floating. A further 18 in. depth of water was pumped into the basin, putting a reverse head on the dam, but still the units would not move from the bottom. After two hours the Agent decided that he must confess to Head Office that for some unaccountable reason these units would not float. Luckily records had been kept showing the rate of ground water recovery after the pumps had been stopped with the first batch of Phoenix units: in the rush to prepare the basin for flooding, the groundwater lowering pumps had not been stopped until the last moment. Although there was ample water around the sides, the units were sitting on a dry bottom. From the original records it was possible to predict when the ground water table would reach the underside of the units and within a few minutes of the predicted time the units bobbed up.

Additional units were also considered necessary for a landing in Belgium as it was thought that this might shorten the war, particularly when the Germans made a stand in the Ardennes. However, the scheme had to be abandoned as there was an acute shortage of tugs and specialist equipment such as pier head pontoons, which could not be manufactured in time.

During all this time work in London was being affected by the V1 flying bombs and the V2 rockets. The site office was demolished by a flying bomb, though luckily nobody was hurt as those working in the basin were below ground level and missed the blast.

ADMIRALTY BARGES

The basin was next prepared for building four deep-draught minesweeping barges for the Admiralty to detonate pressure mines laid in French ports by the retreating Germans. They would be pulled by a shallow draft tug at the end of a long tow line and the heavy bow wave set up by the barge would (it was believed) detonate the minefield. Long before construction was finished other methods had been found to explode mines. In fact, rough seas and high tides had generally been enough.

In March 1945 thoughts were turning to the war in the Far East, as the war in Europe drew to a close. The last use for the Russia Yard basin was to build a prototype steel and concrete caisson to replace the dock gates at Singapore in case these were blown up by the retreating Japanese, thus putting the dock out of action.

Mowlem

Towing out Admiralty barges, Russia Basin

Concrete caisson under construction

Singapore was evacuated so quickly, however, that the Japanese had no chance to sabotage the port. The prototype caisson when completed was towed to Portsmouth and used eventually in No. 4 dock there.

PLUTO AT TILBURY

At the same time as the Mulberry Harbour was being built, there was great activity in Tilbury Dock. Early in 1943 experiments were being carried out on pumping petrol through thin flexible steel pipes, as it was realized that the success of the invasion would depend on keeping a fuel-hungry army supplied with petrol. The initial experiment between the mainland and the Isle of Wight using a 2 in. dia. steel pipeline proved to be highly successful. On 4 April 1943 a further test was carried out between Swansea and the Devon coast using a pressure of 1500 lb/ sq. in. Later the pipe diameter was increased to 3 in. and a total of 710 miles was assembled, of which 140 miles were sent from America. This line, known as PLUTO, was to be laid between the Isle of Wight and Cherbourg.

Another type of pipe, HAMEL, was also developed. This was 2¼ inches in diameter and could be wound onto large cotton reels nicknamed Conundrums. The undeveloped land to the north of Tilbury Dock was found to be suitable for the project and two factories were built for the assembly of 40 ft sections of steel pipe into 4000 ft lengths. There were seven lines of machines with 'flash' welding sets and other equipment. The 4000 ft lengths were laid out to the jetty in the main dock where a further welding set was used to join the long lengths into a continuous pipeline.

The Conundrums, 90 ft wide and 40 ft in diameter, had flanges 52 ft in diameter with teeth for chain-driving the drum. The pipeline, fed by an overhead gantry on to the rotating drum was ballasted so that on the start of winding it was well down in the water. As loading proceeded water was removed to keep it buoyant. A total of 70 miles weighing 1600 tons was held on each Conundrum, and a speed of 7 knots was achieved when laying. In all 17 lines with a total length of 500 miles were laid between Dungeness and Boulogne from the Tilbury plant.

Post-war problems and
reconstruction, 1946–1955 and 1963–1967

REPAIR PLANS

As the war in Europe was nearly over, it was time to consider the havoc and destruction caused by five years of enemy action to the dock installations, and to plan their repair and renewal.

Of 10 500 V1 flying bombs launched against London 2400 reached their target, but nearly all the 1115 V2 rockets fell in the London area. In all 1½ million houses had been destroyed or badly damaged.

When the war in Europe was over, 30% of transit sheds in the Port of London had been destroyed. In the London and St Katharine Docks the figure was over 50%, though other warehouses were capable of being repaired.

Millwall entrance lock had been put out of action in 1940, and the bascule bridge taking the North Circular Road over the King George V entrance lock had been destroyed by a V2 rocket. The entrance jetties had suffered from years of wartime usage and lack of maintenance.

Mainly because so many houses had been damaged or totally destroyed by enemy action, successive Governments gave priority to first aid and permanent repairs to housing. No priority was given to the reinstatement of dock property, and many acres remained devastated more than seven years after the war.

The PLA Board decided first that arrears of maintenance should be undertaken to the full capacity of resources available, and without financial limit. Secondly, partially damaged buildings should be repaired and put back into service as soon as possible. Thirdly, restoration involving extensive rebuilding should be undertaken. This policy was aimed at getting the highest production from the quays, sheds and warehouses left to the Port, before the limited staff and resources were allocated to larger schemes of redevelopment. It would also give sufficient time to consider the post-war shipping requirements.

No one was quite sure how trade would develop and to what extent mechanical handling would be introduced. The need to turn ships round quickly as a means of cutting costs was recognized, but labour-saving cargo handling methods would have to be introduced in stages, and with the co-operation of the dock workers.

SHORTAGE OF STORAGE SPACE

As trade began to return to peacetime levels, the shortage of warehousing became acute. A delivery of 50 000 tons of sugar to a nation starved of sugar for years was most welcome but as no sheds were available it was stored in the open under tarpaulins on vacant bombed sites. At south quay London Dock, where a warehouse had been cleared to quay level, the vaults below were unable to withstand the suddenly imposed load of thousands of tons of sugar and quietly collapsed, depositing the cargo into the basement below. Fortunately the sugar remained undamaged as it was in sacks.

To ease this acute shortage, the Government made available six wartime Marston Sheds. These comprised a light steel framework with asbestos sheets for cladding to the roof and sides. A gang of Italian prisoners of war 'helped' with the concreting of the floors, but untrained and unused to the work as they were, it was difficult to get them to achieve the high standards of workmanship required.

Warehouses in London and St Katharine docks were repaired. The large B warehouse in London Dock had been completely destroyed and was replaced by a long transit shed. The three-storey modern warehouses in Surrey and at the south side of Victoria Dock completed only a year or two before the war had been bombed but withstood the impact well, and only needed repair.

An extensive programme of painting and rebuilding of entrance jetties was put in hand and by 1951 it was possible to contemplate a programme of new works.

With so little secure storage for imports and severe shortages in the country at large, pilfering for sale on the black market was a constant problem. Brandy was smuggled out in bicycle frames, whisky and nylon stockings were wafted over fences, and once 35 cars for export were discovered to be devoid of spare wheels. However, these were not the only problems.

DAMAGE TO JETTIES AND DOCK INSTALLATIONS

In the early 1950s much damage was caused to the entrance jetties by ships attempting to enter the docks. Care would be taken by a pilot when berthing against a solid masonry dock wall as the ship was likely to sustain more damage than the wall! They had little respect, however, for the more flexible timber lead-in jetties as they could berth against them quite heavily without damage to the ship.

When entering a lock after a long voyage on the high seas an entrance appears very small from the ship's bridge. Navigation is difficult, especially on a strong flood tide, and a mistake in control of a large diesel engine can be catastrophic.

In the Shadwell entrance to London Dock a ship going too fast dropped its anchor as an emergency brake. The anchor caught the gate sill, effectively stopping the ship but also seriously damaging the sill and making the lock gates inoperable. As this was the only entrance to London Dock, and many ships were berthed there at the time, it was necessary to carry out difficult repairs at weekends.

The entrance lock, designed by James Meadows Rendel in 1854, was 350 ft long, 60 ft wide and 28 ft deep. Fortunately emergency gate stops had been incorporated in the construction, and after 100 years were still in perfect condition. An emergency gate of heavy steel beams and timber was quickly fabricated to fit the stops, and placed in position by PLA wreck lighters. The lock was dewatered and the extent of the damage exposed. Four weekend occupations were required to carry out the repairs and restore the lock to shipping. The lead-in jetties to the Tilbury entrance lock have been frequently damaged by shipping negotiating the turn into the lock. A ship coming up the river in poor visibility at night mistakenly identified the lights of the cargo jetty for those at the entrance lock. Extensive damage was caused both to the ship and the jetty before the mistake was realized.

Accidents also occur in the enclosed docks as well as the river. The *City of Carlisle* whilst berthing in Tilbury Dock caught the flare of her bows on one of the quay side cranes and the impact force was such that the rail clamp was broken, the quay side crane rails uprooted and the crane turned over on to No. 26 shed behind. The falling crane demolished the shed causing much damage. On another occasion the *M.V. Vogelsand* was carrying a considerable deck cargo of timber which slipped. In an over-enthusiastic attempt to adjust the trim some open portholes were submerged and the list suddenly increased by an alarming amount. Only the timely assistance of the fire brigade managed to avert a disaster.

LOCK GATES
In 1948 following the Chief Engineer's report on the condition of the river entrance and lock gates, the Port of London Authority approved a four-year programme for

Bomb damage, Millwall Entrance Lock

Mowlem

carrying out repairs which had been deferred during World War II. Under normal circumstances it is considered advisable to overhaul lock gates at intervals of not longer than 15 years, but by the end of the war many had been in continuous service for much longer periods. In 1932 60 pairs of gates were in use, but Millwall entrance lock was closed owing to bombing and at the Wapping entrance the old oak timber gates had been strutted in the closed position in 1940 as a safety precaution, so there were only 56 pairs operating in 1949. Between 1949 and 1953, 20 pairs of gates were overhauled and the construction of 11 new pairs was put in hand.

It was decided to maintain three pairs of lock gates in operation at all the important entrance locks, the middle gates being regarded as spares for emergency use. However, in the Royal group of docks, at the time, there could be as many as 50 large cargo vessels berthed at one time. During the reconstruction of the Gallions Entrance lock to Albert Dock, the King George V lock was the only entrance for ships to the whole group. As a precaution against the heavy trade losses which would be incurred in the event of serious damage to the gates, it was decided to keep a fourth pair of emergency gates stored in the vertical position with the timber-work boxed-in at the head of the King George V Dry Dock. Apart from the repair and renewal of the dock gates, it was found necessary to remove and replace pintles, and in a number of cases to redress the hollow granite quoins. To enable this work to be done, two limpet dams were constructed. A large all-welded steel

South Dock Entrance Jetty, West India Docks, after damage

Mowlem

Rebuilding sill, Shadwell Entrance

sectional dam, with handed sill sections, was constructed to be used in the two largest locks at Tilbury and King George V Docks. This was handled by the London Mammoth, or another of the Authority's floating cranes. The second limpet was of composite steel and timber construction, mainly for use in Surrey Dock. This was built of 12 in. by 6 in. timbers bolted to specially fabricated steel joints with a detachable bottom section which could be changed to suit the particular shape of the quoin being worked upon. The seal between the dam and brickwork or masonry of the lock wall was formed by a sausage of canvas packed with oakum. The secret of setting the limpet dams and making them watertight was to establish as quickly as possible a difference in head of water between the outside and inside. The greater the depth dewatered, the tighter the oakum sausage was squeezed against the wall, making a perfect seal. Usually two 6 in. Flight electric submersible pumps were used. With practice the team could set and dewater a limpet in the comparatively short time of 1½ hours. Once the limpet had been dewatered, it could be kept dry by one hydraulic ejector of about 15 gallons per minute capacity.

DIVING BELL

The entrance to St Katharine Dock had been provided with timber sills fixed to masonry blocks. A diver's survey showed that part of the masonry in the sill plat-

PLA

City of Carlisle damage, Tilbury

form was missing completely, and there was an uninterrupted passageway for water under the lock gates. The sill timbers had also been forced out of their position and the steel cover plate had become detached. It was decided to recommission the Authority's diving bell used in 1937 to deepen the Connaught Road cutting, so that emergency repairs could be carried out without closing the lock to shipping.

As barges and small coasters used the lock on the top of the tide it was necessary to find a method of handling the diving bell at low tide when the lock was free. The *M.V. Ebury*, a one-time German seaplane rescue ship, and the pride of Mowlem's fleet, was employed to handle the equipment. Being self-propelled it could go out into the river when the lock was in use. An 18-ton capacity travelling crane handled the 14-ton bell extremely well.

The Author, together with Mr Noel Ordman, then one of the PLA Engineers and latterly a Board member, made the first inspection of the lock bottom. The inspection confirmed the diver's report that the sill timbers were in good condition but displaced. The weight of the bell was used to force them into position and additional bolts were concreted in, the missing stones being replaced by concrete taken down in the bell and placed under water. The whole operation of repairing the damage was completed in four weeks, and provided such a popular spectacle that a special Pathé Gazette film was made of the operation. The star performer was the *Ebury* Captain's dog who was fascinated by the bell disappearing under the water. After this episode, the diving bell was used on many occasions, for example the deepening of Garnet Street cutting in London Dock, sill repairs to the Shadwell entrance lock, the removal of the limpet section of dock wall in the Connaught Road cutting, and setting the sill blocks for the emergency flood gates at Tilbury Dock. On three separate occasions it was sent up to Liverpool to help with emergency repairs.

RECONSTRUCTION OF GALLIONS ENTRANCE LOCK

The Gallions entrance lock had been built by the Engineering Department of the London and St Katharine Dock Company using direct labour. It was based on the Gallions Upper Entrance or barge lock designed by Sir Alexander Rendel, and had been opened to traffic on the 20 July 1886.

The length and beam of the lock, 550 ft by 80 ft, were in reasonable proportions and adequate to take up to two-thirds of the ships using the Royal group of docks. For many years it had taken a vast quantity of shipping and had been used as a ship lock up to the outbreak of World War II. Its condition had seriously deteriorated, and had it not been for the war, it would probably have been reconstructed at an earlier date. However, two severe restrictions made it essential to undertake the works. The old gates were inconvenient, being chain-operated and seventy years old; more important, the lock bottom was curved to a radius of 110 ft. This was no

M. V. Vogelsand, Surrey Dock

69

Limpet, ready for lowering into position   *PLA*

doubt suitable for Victorian ships, but modern vessels are practically flat-bottomed. The proposal was to reshape the bottom by cutting away the curved haunches, substituting reinforced concrete knees. All granite work was to be reconditioned and the quoins redressed, and damaged and loose brickwork facing of the lock walls would be replaced by granite aggregate concrete fixed to the concrete wall by anchor bolts. Modern all-welded buoyant gates operated by electrically powered machinery would be installed.

Boreholes put down through the bottom of the lock confirmed the presence of about 10 ft of Thames sand and gravel overlying fissured chalk. They also showed that the thickness of brickwork and concrete in the floor appeared to be less than old records had indicated.

Competitive tenders were called for, and a contract was awarded to John Mowlem and Company Limited on 25 September 1951, although the lock was not

70

closed to traffic until 1 January 1952. The contractor was called upon to provide a system of ground water lowering to prevent uplift during reconstruction and as an additional safeguard in the event of power failure, the base of the lock was to be loaded with 10 tons of kentledge per foot run prior to dewatering.

The choice of dam construction was left to the contractor, but safety to shipping in the whole Royal group of docks was of paramount importance. A dam failure caused by the collision of a ship turning in the Albert Dock basin would have been catastrophic. The Port Authority called for unlimited third party indemnity, but the underwriters would only accept a liability up to £3 million. In view of this it was decided to use gravity block dams, with joggle jointed interlocking blocks, each weighing about 11 tons.

The inner dam was founded on a prepared concrete base levelled by divers, the concrete being placed under water using tremie gear. The apron to the inner end of the lock was extended to form a strut between the base of the dam and the body of the lock while the dam foundation was being constructed and eliminated all problems of sliding.

Mowlem

M. V. Ebury with diving bell, St Katharine Dock

71

This type of dam had been used successfully on three separate occasions previously, at the new entrance to Tranmere Dock, the entrance to Victoria Dock, Hull, and the Dover Train Ferry Dock. The secret in this type of construction is to pay special care to the accurate setting of the first row of blocks. If this is done correctly the remainder fit together without difficulty. The inner dam was 48 ft high and the depth of water 42 ft. The overall length was 207 ft, and the weight 15 000 tons. Steel piled dolphins with half timber fendering were provided as a protection against ship damage.

Mowlem

Gallions inner blockwork dam

When the dam was being tested, movement took place between six of the rows of blocks. While horizontal movement was arrested by the interlocking recesses cast in the tops and bottoms of the blocks, a survey showed that a displacement of about one inch had taken place between adjacent courses. From calculations carried out, the angle of friction must have been reduced from an expected 30° to 17½° for sliding to have occurred. The reduction was probably due to the presence of silt mixed with oil which continually settled on the blocks already laid. No amount of cleaning would entirely remove this mixture, and its fine greasy nature acted as a lubricant between the courses of blocks. No further movement occurred once the tolerance had been taken up in the joints, and leakage through the dam was quite small.

On surveying the foundations for the outer dam, it was found that a scour hole some 20 ft deep had occurred in the chalk, beyond the lock apron. This made it extremely difficult to construct the outer dam in blockwork as anticipated.

After some consideration it was decided to abandon the idea of blockwork and to use a double skin sheet-piled gravity dam. The two skins were 40 ft apart and the top of the dam 18 in. above the level of the highest known tide. The 70 ft long piles penetrated approximately 15 ft into the chalk. The filling between the sheet piles and on the berms each side was of brick and concrete rubble from London and St Katharine Docks, plus a large quantity of granite sets. The total volume of hardcore employed was 30 000 cubic yards. When the lock was dewatered and the full head of high water spring tides was taken on the dam, a deflexion of 14 in. was recorded at the centre, about what would be expected in a dam of this type.

The PLA found many uses for the Gallions blocks on the completion of the contract: they were used for a permanent dam to seal off the damaged Millwall Dock entrance; blockwork facing was used in the infilling at Wapping entrance; some blocks were used as kentledge in St Katharine entrance lock during its reconstruction in 1958, and others were used to face a dam constructed across the old entrance to Tilbury Dock. A similar type of block but of larger size was used for carrying out emergency repairs to Newport Dock some twenty years later.

A ground water lowering system comprising 34 deep wells surrounding the lock was installed to reduce the hydrostatic uplift on the floor. This also reduced the pressure on the walls and in the dams, to reduce uplift and permit the full development of sliding resistance. The 24 in. dia. wells were sunk into the chalk and provided with 12 hp submersible pumps. The ground water table, initially at +2.0 OD was reduced to and maintained at a level of −28.0 OD during reconstruction.

The permanent works including the reshaping of the invert were carried out without incident and the lock was reopened to shipping in 1956.

WIDENING THE CONNAUGHT ROAD PASSAGE

The second major civil engineering construction in the Royal Docks was the widening of the Connaught Road cutting. As described earlier, this had initially been deepened in 1937 to allow ships of an increased draught to be berthed in the Victoria Dock.

In 1958 the Royal Mail Line had under construction three new fast refrigeration ships for the South America meat trade, which usually docked at Z berth in Victoria Dock. The beam of the largest of these, the *M.V. Amazon*, was 78 ft 2 in. and the cutting was only 84 ft wide, leaving little clearance. It was decided to widen the cutting to 100 ft by rebuilding the wall on the south side except for 40 ft at the abutment to the swing bridge. Provision was to be made, however, for a clear width of 100 ft throughout, should the construction of a new bridge be authorized at some future date.

The new wall was constructed behind the old wall using Larssen sheet steel piles tied back to a piled anchor beam. Over the tunnel section where piles could not be

73

driven a sloping re-entrant would be formed and decked over to provide continuous coping.

Companies tendering for the work were given the opportunity of choice of method for the demolition of the old wall. Should the use of explosives be chosen then the cutting would have to be swept clear of debris after each round of charges had been fired. A guarantee that all rubble had been removed had to be given before the passage could be reopened to shipping. The risk of a block of concrete being missed in the sweep and getting jammed between the hull of a ship and the roof of the tunnel was considered to be too great by the successful tenderer John Mowlem & Company Limited, who chose instead the risky task of demolishing the old wall in the dry inside a cofferdam. The PLA allowed a maximum encroachment of three feet into the passage for temporary works, just sufficient to drive a row of 4 B Larssen piles in front of the wall to form a cofferdam. Where the British Railways tunnels passed under the dock wall it was necessary to construct the cofferdam in the form of a limpet. The toes of the piles were bolted to the wall and

Aerial view of Connaught Road Cutting

Derby Aerosurveys

PLA

View of 200 ft span shed, Victoria Dock

concreted in by divers to form a watertight seal. A heavy fender beam, supported on the cofferdam piles, provided protection against passing ships, with the thrust from a blow taken back to a 'deadman' built into the coping of the new wall. This fender system worked successfully. It provided protection to the dam and at the same time gave uninterrupted passage for shipping. During the six months that the dam was in position, 360 ships of varying sizes, and 1200 tugs with innumerable barges passed through the cutting without delay. The passage of these craft caused 94 incidents, ranging from slight abrasions of the fendering to the occasion when a ship laid heavily against it, buckling the cofferdam 16 ft below the surface. Although the piles took up a deflexion of about 10 in. the cofferdam remained intact until the very end.

The 5000 cubic yards of brickwork and concrete from the old wall was split into blocks of about two cubic yard size and lifted out by derrick. When the bottom had been almost reached, four filter wells were put down into the Thanet sand below the Woolwich and Reading beds to relieve artesian pressure. The flow from these wells was small.

A cut was next taken through the back of the old wall to expose the crowns of the railway tunnels some 9 in. below. These had been lined with steel rings but the gap between the rings and the old brickwork had to be hastily filled and a watertight seal made with concrete. When the tunnel had been made safe the remaining section of the wall was thinned down until a piece 7 ft wide remained supporting the toes of the limpet section piles. This was considered to be of sufficient strength to provide a safety margin, but on a Sunday morning the brickwork began to give way, allowing water to flow into the works. Extra pumps were hurriedly brought to the site and these kept the water down for the remaining 200 cu. yd to be prepared for demolition under water.

Temporary timber struts were placed alongside the steel struts in the cofferdam. As flooding took place in stages, the steel struts were removed, and a few days later when the water inside had reached the level of that outside, the timber struts floated to the surface. About 180 tons of steelwork were thus removed without the aid of a diver.

Whilst the widening of Connaught Road cutting was in progress, it was decided to extend the new piling into Victoria Dock in the form of a return quay. This would join on to the last monolith of the 1936 construction: dredging away the intervening peninsular would complete the whole Victoria Dock complex, providing two additional ship berths. Construction of the new wall in the region of the old southwest knuckle of the cutting was difficult. The line of sheet steel piling intercepted a jumble of over 100 timber and steel piles with their associated tie rods, bracings and anchor blocks which had to be removed. Work was made still more complicated by thousands of broken earthenware jars which had been tipped in the area after enemy bombing in 1940 when a warehouse containing much of the nation's breakfast marmalade had been destroyed. The flow of water through the broken earthenware prevented effective dewatering, and concrete was placed by tremie pipe in this section of the quay. The south side provided accommodation for two ships, each 470 ft long. These were to be served by a transit shed 696 ft long with a clear span of 200 ft, giving a floor area of nearly 140 000 sq. ft. Since the cost of providing a single-span shed had been found to be very close to that of a double-span shed of the same width, it was accepted that internal columns should be avoided. This was to be particularly important with the introduction of mechanical plant for cargo handling.

Covered end bays for loading lorries in the dry were provided with a clear headroom of 31 ft. For such a large span shed it was decided to adopt a three-pinned arch with ties, rather than a truss. The roof outline was made approximately parabolic, but was kept rectilinear for ease of fabrication. There was a minimum height of 20 ft to the doors and ties, the arch rising another 35 ft above this level. As the trusses were 25 ft apart, there was insufficient room to store cargo more than 20 ft high, a fact recognized in the design of later large-span sheds which used monopitch roofs with cantilever columns.

RECONSTRUCTION OF ST KATHARINE ENTRANCE LOCK

A unique opportunity occurred to compare the engineering works of Thomas Telford with those of G. P. Bidder when the PLA found it necessary to reconstruct the St Katharine Entrance Lock in 1957, and the Western Entrance to Victoria Dock in 1963. The St Katharine entrance was built in 1827 and lasted 130 years. It had been constructed in the very best traditions of the canal age using good quality workmanship and masonry. The Western Entrance built at the height of the Railway boom used the latest materials then available, cast iron piles and plates, and lasted just 100 years with the slimmest of safety margins.

Although emergency repairs at St Katharine's had been carried out in 1952 by diving bell, it soon became apparent that the problem of the gate platforms to the entrance lock was deep-seated. Whenever the lock gates were closed, water boiled up through the platform indicating that there was a free water course under the gates. The situation became so bad that by 1957 it was found necessary to close the dock completely to barge traffic, and rebuild the gate platforms. Tea, the main com-

Mowlem

Water boiling up under gate platform, St Katharine Dock

modity using the Mint Warehouses on the wharf, was unloaded at the nearby Orient Wharf and ferried across the road to the St Katharine Dock.

It was decided to install completely new reinforced concrete gate platforms and sills with steel sheet piled cut-offs driven into the London Clay below. As the lock would only be used by barges and small lighters on the top of the tide it was found possible to install bottom hinged flap gates using steel cable raising gear with winding winches.

The inner dam comprised a single skin of steel piles driven into the dock bottom, supported by three 6 ft deep plate girders abutting against special bearing blocks fixed to the dock wall. Since the river end of the lock walls would have to be cut back to accommodate the new gates it was essential to keep all pressure off them and a double skin steel piled gravity dam was employed. As in the case of the outer dam at Gallions it was filled with brick hardcore, and provided with observation wells large enough to take electric submersible pumps in an emergency. On testing the dam, the deflexion at the centre was 10 in. Before dewatering it was considered prudent to weigh down the curved barrel of the lock to prevent uplift, and as a further precaution holes were drilled through the brick invert to relieve the pressure from any water which might be found beneath.

When dewatering had been completed, there was nothing but admiration for the

J. Flowerday

Missing stones from gate platform, St Katharine Dock

excellent state of the masonry and brickwork which had been under water for 130 years. Each block of the Bramley Fall stone was beautifully tooled, and the brickwork was in good condition. The original oak gates had been made in the best tradition of the shipbuilder's art. Bronze fixings and spikes had been used throughout to prevent corrosion. The top sections of the gates between wind and water were badly decayed by weather and abrasion from barges, but the lower parts, constructed of close-jointed 15 in. square timbers, had kept their strength. The heel posts were massive, 18 in. by 20 in. whole timbers 33 ft long, and the mitre timbers were 18 in. square. All these timbers were later sold to a merchant specializing in the supply of timber for restoring old houses. It is hoped that after 130 years as dock gates they may have found a home in some country house for a further period.

78

The reason for the failure of the gate platforms soon became apparent, however. Masonry blocks had been laid on timbers supported below by a framework of whole timbers on piles. The space between had been filled with a gravelly mixture of stones and puddle clay, which it is believed was used by Telford as a mat to prevent the carpenters disturbing and 'puddling up' the London Clay in the bottom whilst fixing the platform timbers. With the shaking up that the lock received during the bombing of the nearby warehouses in the war, water must have found its way under the timbers through the gravelly mixture with sufficient force to displace the stones above. All the timbers were in good condition.

Western Entrance, Victoria Dock, demolition of wall

Nevertheless, there must be one black mark against the masons employed by the Victorians. One afternoon, by coincidence during a Director's visit with the Agent and other members of the staff on parade, and whilst they were watching the cutting back of masonry, a stone 7 ft long, 18 in. high, but only 1 ft deep suddenly blew out of the face. Water, gravel and sand rushed in to the lock followed by mud and rubble. Before the flow could be stopped the ground surface behind the wall caved in leaving a large cavity. It is seldom, if ever, that a mishap of this nature is laid on for the Directors, especially as the mason 130 years previously must have thought he had 'got away with it'.

Clear water flowed in at a rate of 40 gallons per minute and first aid repairs were carried out by packing 12 in. by 12 in. timbers over the area covered by the missing stone. The cavity behind was filled with ashes, and steel sheet piles were driven around the back of the wall to form a limpet seal. This stopped the flow of water and the timbers could be removed, the space of the missing stone bricked up and cement grout injected behind to form a permanent seal.

RECONSTRUCTION OF THE WESTERN ENTRANCE LOCK TO VICTORIA DOCK, 1963–1967

The Western Entrance lock of Victoria Dock was built between 1853 and 1858, and was considered to be the most advanced design of its time.

The finger piers inside the dock had been constructed using cast iron T piles, with cast iron panels between, and the main body of the lock was of similar construction. This type of piling was similar to that successfully employed in the construction of Brunswick Wharf, Blackwall, and at Fleetwood Harbour. The cast iron piling formed bays 7 ft 1 in. centre to centre of the main piles, the space between being filled with cast iron plates retained by the flanges of the main piles. The piles were 37 ft 8 in. long with a front flange 18 in. wide, the metal being 2 in. thick. As it was not possible to cast, transport, and drive such large piles, they had been pitched in two halves, and joined with splice plates and bolted connections.

The main piles were tied back to a row of timber anchor piles driven only 18 ft behind. Two 2 in. dia. tie rods were used to each pile, one attached to the top of the lower section of the pile, and the other to the top section of the pile some 8 ft above the splice. The space between both rows of piles was filled by a stepped concrete wall of lime concrete, the proportions of which were six parts of Thames sand and gravel to one part of 'Halling' lime. The lime was not ground, but used hot, and in the words of the Resident Engineer of the time 'was found to set very hard'.

Unlike those at the Gallions entrance lock, the strata were simple, clay and peat (the clay used for bricks, and the peat sold as fuel) and sand and gravel overlying London Clay at −25.0 OD. The ground water table stood at about Ordnance Datum level before work started. Ground water lowering had been carried out by two 'great pumps' lifting water from sumps on the north side of the lock just behind the line of the wall, and between the gate positions. During the whole of the time that work was in progress, the surrounding countryside was gradually drained of its water for a considerable distance. Two and a half miles away in East Ham the level of one well dropped for the whole of the time that the pumps were in operation. The gravel was completely drained down to the London clay, which on being exposed made an excellent foundation for the gate platforms. Platforms 120 ft wide and 73 ft long were laid directly on the London Clay and around each a single row of elm sheet piles, 16 ft long and 8 in. thick, was driven to a depth of 7 ft into the solid clay, the tops being neatly finished off, and capped by a waling. Inside these stockades, 8 ft 6 in. thick platforms and the quoin buttresses were built in massive brickwork laid in 2:1 lime: mortar, again using Halling lime. The quoins and other stonework facings were of a compact sandstone called Duke's Quarry stone. The gate platforms were in close contact with the London Clay and the problem of leakage under the gates did not arise. In the body of the lock the gravel was removed and replaced by puddle clay, a concrete bottom not being considered necessary.

Before the lock was completed, two failures occurred. The first was caused by differential settlement between quoin buttresses and the middle of the sill. Though only just over 1 in. this was enough to fracture the cast iron sill about 3 ft from the mitre end. The Engineer admitted later that he should have foreseen this before allowing the sill castings to be bolted down.

Excavation at Western Entrance, Victoria Dock showing 1928 gate platform and previous wall, on London Clay

The second failure was almost catastrophic. On Sunday morning 17 June 1855, when the gates had been installed prior to flooding, backfilling behind the walls was nearing completion, and pumping had stopped for some weeks, the full hydrostatic head of water came on the north and south walls together. Without warning a section of the north wall between the gate platforms moved forward bodily into the lock, pushing up the thick puddle clay towards the centre. The tie rods were bent and broken and the anchor piles dragged forward and in some cases broken. A few hours later the south side wall failed in the same way, though the platforms and quoin walls remained unaffected. The Resident Engineer admitted that the substitution of clay puddle for the gravel in the bottom was, to a certain extent, an evil, as the walls were thereby deprived of weight at the foot, the ten-

Mowlem

Reconstruction nearing completion, Western Entrance

dency of which would have been to prevent forward motion. This of course was correct, but in addition the anchor piles were inadequate, and too close to the wall. As will be seen later, the concrete made with lumps of hot lime in fact had little strength, certainly insufficient to resist the movement of the wall.

The most energetic measures were adopted for reinstating the walls. The old concrete was cleared away. The piles were withdrawn and redriven on the original line, but to 10 ft greater penetration into the London Clay. A solid concrete wall was founded behind the piles, 3 ft into the clay, and carried up to 18 ft, with a base thickness of 18 ft. Above this level the thickness was reduced to 10 ft with counterforts 6 ft deep. The piles were tied back to the concrete by rods passing through it, secured to cast iron plates tight against the back of the wall. The whole of the puddle clay was removed, and replaced with lime concrete topped with a 6 in. layer of Portland cement concrete. The whole of the reinstatement was carried out in little more than three months, an extremely short time remembering that the pumps would have to be reinstalled, and the ground water table again lowered, before rebuilding could commence.

Later there was little trace of the concrete bottom. This had no doubt been taken out by dredgers over the years when carrying out routine maintenance mud clearance.

In spite of a number of minor accidents to the gates and sills, the lock worked satisfactorily for 70 years. By 1928 the gates had reached the end of their useful life,

82

and the roller paths were giving trouble. It was decided that the Western Entrance should be used solely for barge traffic as by now all shipping was using the Gallions and King George V entrances, and London County Council were able to go ahead with plans to build Silvertown Way with a fixed span bridge over the entrance lock. A fresh platform was built on top of the existing outer gate platform to strengthen the gates and roller paths. This reduced the depth by five feet but this restriction was acceptable as only barges would be using the lock.

Construction of the new platform was carried out in concrete by using a half-tide dam built in timbers and new gates were provided to the shallower depth.

These repairs lasted for nearly 30 years, when more problems began to arise. Several of the cast iron panels broke, allowing the rise and fall of the water in the lock to leach out the lime from the concrete backing. In the outer lower side bell-mouth the problem became acute, as the gravel concrete disintegrated and ran out into the river through gaps in the broken plates, leaving large holes in the wall, and emergency sheet piling had to be driven in front of the wall to prevent further collapse. During 1957 leakage of water through the lock increased, requiring excessive use of the impounding pumps to keep the water up to the correct level throughout the Royal group of docks. In November of that year it was discovered that the heel post of one of the gates was becoming detached, and at this point it was decided to close the lock to traffic, and a dam of hardcore and sheet piling was quickly constructed close to Silvertown Way to protect the Royal Docks from massive loss of water.

Financial conditions at the time meant that any major reconditioning would have to be postponed, as more important works had to be given precedence, but when it became known that rebuilding was not going to take place immediately, there was a loud outcry from the lighterage industry. They objected to using the Gallions lock and Woolwich Reach, which not only increased the distance up river by three miles, but meant that they had to negotiate the narrow and extremely busy Albert Dock. In fact 80% of all barge traffic had become accustomed to using the Victoria Dock entrance and did not like having to change route.

It was not until 1963 that authorization was granted for the complete remodelling of the West Entrance lock. During the six years of closure the PLA Engineering department had time to consider various methods of carrying out what was likely to be a difficult task of reconstruction, and when in May 1963 they called for competitive tenders, the design, shown on 120 drawings, had been completed down to the last detail.

The scheme provided for two new pairs of lock gates, in new outer and middle positions. The old outer gates built in 1928, were to be modified to fit the new inner position. An important feature of the project was to be a new deep lock extension on the river side of the old outer gates to enable locking to take place at all states of the tide. The body of the lock was to be supported by heavy plated steel piles driven in front of the toe of the old piles, the upper portion of reinforced concrete being tied back to prestressed anchor piles behind the old wall. It was not envisaged that any work would be necessary to the balancing culverts, as the original drawings showed them to be constructed in brickwork with substantial concrete surrounds.

In the rebuilding of Gallions Entrance lock, ground water lowering had been used successfully and the designers considered that it would again be used.

The tender submitted by John Mowlem & Company Limited contained two important suggestions. The first was substitution of bored piles for driven reinforced concrete piles to support the floor and walls of the lock extension. It was felt that problems arising from vibration, heave and remoulding of the London Clay would have serious effects on the stability of the old structure. The second concerned the dewatering. At Gallions Entrance there was water-bearing chalk under the gravel but here with only a little gravel overlying the London Clay, it would be difficult to obtain sufficient immersion depth in the wells to provide satisfactory dewatering of the gravel. Undoubtedly pumping from open sumps as used by the Victorian engineers would draw fine material from the gravel, adding further to the problems.

Instead of dewatering, it was suggested that a cut-off in the London Clay would be carried out by driving steel sheet piling around the balancing culverts. Where no piles could be driven through the old walls, the cut-off would be completed by chemical consolidation of the gravel.

These alternatives were accepted by the PLA and a contract valued at £1¾ million was awarded with a starting date of 23 September 1963. In order to construct the outer cofferdam from both sides of the existing lock entrance it was necessary to form the new bellmouth walls against which the dam would abut. While preparations were being made to concrete the new outer walls inside the sheet piled cofferdam, the old wall started moving forward, and in fact moved three inches before the weight of new concrete stopped it.

Drillings were put down through the old concrete and showed that this was in an extremely poor condition, and that in many places the lime had been leached out, leaving a gravelly mixture. It was decided to abandon the plans for chemical consolidation, and to drive steel sheet piles through about 35 ft of Victorian concrete. This was achieved without difficulty! At the same time concern was expressed over the state of the balancing culverts. Probings could discern no pressure from the concrete surround to the brickwork. It was not possible to examine the actual state of the culverts until the lock had been dewatered. When it was possible to uncover them it was found that some of the brickwork had crushed and settlement had taken place. Tests carried out by the Building Research Station showed that the shear value of the mortar was of a very low order. There was no concrete under the culverts and in fact they were not constructed as shown on the drawings, and were located in different positions relative to the quoins. As the stability of the gate walls depended on the existing culverts acting as counterweights, it was not possible to remove the gate platforms.

When the culverts had been completely rebuilt the opportunity to incorporate modified top gate anchorages into the new construction was taken. As the quality of the brickwork was so poor, especially below low tide level, it was decided to cut back the face and encase the whole of the gate wall in concrete.

The outer dam was 220 ft long, 50 ft wide, and 46 ft high from formation level to the crest at flood prevention level. It was constructed with three skins of sheet piles, the 25 ft wide space between the inner and middle rows enabling the 7 ft thick apron to the new entrance to be constructed underwater using tremie gear and underwater skips. Precast concrete stop ends were lowered into position under water between the pans of the piles, dividing the area into sections representing one day's concreting. When the apron had been concreted the whole of the dam was filled with brick rubble hardcore. Two observation wells, each large enough to take a submersible pump, were provided to monitor any leakage which might seep through the piles. On testing at high water spring tides the deflexion in the centre of the dam was 14 in. – similar to that at Gallions.

At the inner end of the lock, advantage was taken of the existing emergency dam suitably strengthened by an additional row of sheet steel piles and hardcore. Down each side of the lock the cut-off was completed through the gravel by driving over 3000 tons of sheet steel piles into the London Clay below. These had to be driven in front, behind, and even through the old walls before dewatering could take place. As soon as it was possible to dewater the lock, a number of unexpected obstructions were encountered. One of the most puzzling was a considerable number of wooden barrels filled with hardened cement. Research disclosed that many years before, a barge load of cement had overturned in the entrance, depositing its cargo over the lock bottom. Scour had taken place beyond the outer gate platform, and attempts to arrest this had been made by using steel plates, and large blocks of sandstone. In the body of the lock remains of cast iron and timber piles were found where the slip had occurred during the building of the lock. One of the few safe places upon which the 20-ton derrick cranes could be sited was considered to be the massive brick walls, but when the gate platforms had been uncovered it was found that someone in 1855 had decided to 'bottom up' the excavation and begin the brick-work construction 7 ft higher than shown on the drawings! This led to hasty re-arrangement of the derricks and the driving of extra sheet piles to contain the old foundations.

The new gate quoins presented a particular problem. They were to 15 in. radius, forming a semi-circular recess. At the outer gate positions the quoins were to be 38 ft high with stainless steel castings at top and bottom to take the bearing press-ure. It was decided to precast the quoins in 12 ft lengths each weighing 5 tons. The accuracy called for was $\pm$ 1/32 in. both radially and vertically. To achieve this very high degree of accuracy in reinforced concrete, a 2 ft 6 in. dia. steel pipe was specially turned on a lathe as a former around which the units were cast. The units were positioned and concreted into the dock walls, with 1/16 in. dry joints sub-sequently injected with epoxy grout under pressure.

As soon as the reconstruction had been completed the problems of installing the new gates had to be overcome. The gates were assembled on the crosswall berth in

Albert Dock and weighed up to 120 tons each, measuring 46 ft by 40 ft, and were 6 ft wide at the centre. As it was not possible to enter the lock from the Albert Dock owing to the limited headroom imposed by Silvertown Way, the only method was to carry them to the site by the river. To do this the London Mammoth, the PLA 200-ton capacity floating crane, was used. Each gate was carried to the site, swung over the cofferdam at the full extent of the crane's safe reach and lowered into the lock. The whole process took four days during which the lock had to be filled and emptied on each tide, as the cofferdam had not been designed to stand a reverse head with the lock full of water at low tide.

With the gates installed the official opening was carried out on 13 April 1967 by Lord Simon, the Chairman of the PLA. It was expected to relieve the work load of the two eastern entrances by taking away much of the barge traffic, but the container revolution soon removed shipping from the Royal Docks, wasting all the difficult reconstruction work.

Modernization of cargo handling and dredging problems, 1950–1963

CHANGES IN CARGO HANDLING, AND TRANSPORT, AND DEVELOPMENT OF LARGE SPAN SHEDS

By 1950 two important changes in working conditions had begun to influence the design of new transit sheds, namely the introduction of mechanical handling, and the use of road transport instead of rail.

The effect of mechanization had been to reduce the number of doors in transit sheds to four per side, increase the clear headroom to 20 ft, reduce the number of columns and initially increase the size of loading banks. The increase in road traffic made it essential to widen quays, roads, sheds and alleyways, and to remove all huts and obstructions around the main buildings. Offices for staff, customs and shipping staff were moved to upper floors with gear stores and baggage lock-ups beneath.

The first new warehouses to be built after the war, No. 10 at London Dock, and Nos 10 and 11 in the Import Dock at West India, did their best to follow these principles, but the limitation of the sites did not allow much room for road traffic. Numbers 10 and 11 warehouses were built on the burnt-out site of the old rum quay warehouses. The basements had been filled in with debris from the fire, and heavy cast-iron rails were embedded in the basement floors: these had to be removed prior to piling. The quay repeated the false quay designs used so successfully in Albert Dock in 1937. The warehouses, 432 ft by 128 ft, were on three floors. The ground floors, for transit purposes, were 20 ft high. The two upper floors, equipped with balconies for unloading direct from ships, were 12 ft 6 in. high and used for warehousing. The design was similar to that employed for the five warehouses on the north side of Victoria Dock, and represented the best in warehouse development at the time.

At No. 10 warehouse London Dock, two sections had been completely burnt out, and were reconstructed in a reinforced concrete framework with brick panels, the upper floors corresponding to the remaining portions of the old building. The new ground floor was made 20 ft high by omitting one floor. It was designed to withstand pressures of 4 cwt/sq. ft which enabled mobile cranes and fork lift trucks to be

General view of prestressed concrete framework, New Guinness Warehouse, London Dock

used. These, apart from M Shed in West India Dock, were the last to be built in re-inforced concrete.

In 1951, at the time of the Korean War, there was an acute shortage of steel, par-ticularly reinforcing bars. The PLA had authorized the reconstruction of No. 19 berth London Dock to provide facilities for the short-sea vessel trade. The transit shed was to be suitable for accommodating the large variety of goods in this dock, using mechanical-handling equipment. Ample loading banks were required as the berth was not served by rail, and the restoration of the vaults was important. An ex-periment was made in using prestressed concrete for the shed framework: a steel frame would need 150 tons, a precast concrete frame 36 tons of steel, but a pre-stressed framework required only 3 tons, which would make a substantial saving. The building, 257 ft long and 150 ft wide, was spanned by two 75 ft frames. These were pin-jointed at the apex and at the base of the columns. Each frame consisted of two rafters, and two columns. All sections were post-tensioned using the Freys-sinet system. Cables were threaded through ducts and anchored at the base and apex, and tensioning was applied by jacks at the shoulder joints. The purlins and sheeting rails were precast and prestressed on the long-line system.

A criticism of the design is the complexity of the arrangements of anchorages and mild steel reinforcement necessary to take the concentration of local stresses caused by the eight anchorages. The roof and side sheeting were of aluminium so that the whole shed would be maintenance free, but as London Dock was closed in 1969 this gave only 17 years to test the results.

MECHANICAL TIMBER HANDLING IN SURREY DOCK

In September 1940 350 000 tons of timber in Surrey Dock were burnt in one night of enemy bombing. After the war the reconditioning of 30 million cu. ft of shed space took place, mainly by repairing bomb- and fire-damaged steel-framed sheds. Mobile cranes with cranked cantilevered jibs which could reach over to the back of a stack of timber 20 ft high were now being used, but for these a hard working surface was essential. To gain access to the 150 acres of stacking space provided, either under cover or in the open, 4¼ miles of 10 ft wide concrete alley-ways were constructed.

The old deal porters who over the years had developed the art of carrying a plank ashore down narrow runways constructed from the deals themselves, bouncing along in such a way that the spring in their stride enabled them to run up stacks of timber sometimes 50 ft high, were no longer needed. In spite of the introduction of mobile cranes to handle the timber and speed up unloading, every berth was occupied during the busy months of July and August. Ships had to queue up off Southend awaiting their turn. This state of affairs lasted for about ten years from the mid-1950s when Surrey Dock was at its peak of success.

By 1963 thoughts were turning to the packaging of timber in unit loads. There were many problems to be solved, such as the weight of the unit, the condition of the wood, transportation and selling. To be economic, timber needs to be packaged according to length, with consistent height and width throughout a shipment, and the package has to be banded in two or three places to withstand handling. Once a package had been made up, the tasks of loading and unloading became much easier. The handling of timber, always regarded as one of the most difficult of cargoes, would be greatly simplified. The Timber Research and Development Association set out as requirements for successful packaging:

Stacking cheese in new vaults, Guinness Yard

Mowlem

Interior of new shed, Millwall Dock development

(a) deep water berths to take sizeable bulk carriers
(b) clear dock sides and good service roads
(c) sizeable back-up areas to take a complete shipment of timber, preferably close
   to the quay.

These requirements, when adopted, spelt doom to the long established Surrey Dock. A hotchpotch of small timber ponds and shallow docks which had grown up over 150 years, mainly centred on the small ships used in the Baltic timber trade, could never accept large bulk carriers. Closure came in 1970 as package timber was adopted. The trade was transferred to Tilbury which soon became the major timber port for large unit load carriers.

MILLWALL DOCK REDEVELOPMENT
The fork-lift truck had been introduced to Britain by the American forces during World War II and its innovation was surrounded by a certain aura from the performance produced in the hands of the Americans who occupied some of the docks during the war.

In order to gain the full advantage from the fork-lift truck, however, cargo has to be received at the point of unloading on pallets. These would have to be of a standard size and travel with the goods to the delivery point which might often be in another country. Ideally ships should be designed with side-opening doors to enable goods to be loaded. The universal adoption of pallets was very slow and new consignments, either import or export, were received in unit load form. The PLA, determined to make palletization a success, was forced to do the next best thing – palletizing the cargo direct from road or rail vehicles. The pallets could then be handled as unit loads all the time that they were in the Port Authority's custody. This method reduced the ship turn-round time by 20% and became popular with the shipping companies.

Because the PLA was strongly in favour of the development of the new techniques the bold step of redeveloping the whole of the Millwall Docks was taken in 1957. The old low brick sheds were demolished in turn, to be replaced by modern sheds with 150 ft clear spans and over 20 ft height to the roof trusses and doorways, ideal for the modern cargo handling methods. The development would have to be carried out over a period of ten years as only one berth could be released at a time.

One of the major obstacles to the PLA's planned layout of the area was Glengall Grove. This road, which ran between the inner and outer docks at Millwall, was flanked by high customs walls, and connected the east and west sides of the Isle of Dogs via a drawbridge 80 ft wide constructed by the Millwall Dock Company in 1867. The knuckles of the old drawbridge restricted the best arrangement of berths

Traffic holdup at Glengall Grove c1928

PLA

but a replacement for the bridge was still needed, even though since 1945 it had taken only pedestrian traffic. After considering a tunnel under the dock, which would have had to be expensively constructed using compressed air, it was decided to retain the knuckles of the drawbridge and to follow the line of the old road with an elevated walkway. This would have a clear height of 30 ft above the dock to enable barges, tugs and small craft to pass underneath without lifting the centre span.

The final design agreed in 1962 provided for a covered bridge on one level and in a straight line with lifts and stairs at each end, and suspended from the roofs of F and G sheds so as not to restrict working space. The opening section of the bridge was provided by a single bascule 113 ft long, which at night provided an illuminated landmark 140 ft high when the bridge was open. The machinery and counterweight were concealed in a reinforced concrete tower situated on the west knuckle.

EAST SIDE OF MILLWALL DOCK

The Fred Olsen Shipping Line were strong advocates of the fork-lift truck and pallets as the most economical method of handling certain cargoes, particularly tomatoes, and co-operated with the PLA in the layout of three new berths, Nos 1, 2 and 3 on the east side of Millwall Dock. Designed to accommodate two new 9500 ton express fruit/passenger vessels, the *Black Watch* and *Black Prince*, the berths were intended to provide the first high throughput palletized unit cargo terminal in the world.

Each berth was provided with a wide quay free of cranes and rail tracks. The sheds were 200 ft clear span and 600 ft long with the floor space completely uninterrupted. Ample back-up space was provided by reclaiming an area of waste ground behind the sheds. With the berth fully manned, it was possible to turn a ship round in six hours. The ships had side-door loading so that fork-lift trucks could work in the holds as well as on the quayside.

Following the modernization of Millwall Dock, the South Dock of the West India complex was improved by demolishing the old L and M sheds and building a new M shed on three floors. The ground and first floors had a cargo capacity of 163 000 sq. ft, the second floor 78 000 sq. ft. Road access was provided to the first floor so that more intensive use could be made of the berth and supporting areas. Import vessels by discharging into alternate floors of the shed were able to follow each other into the shed in quick succession. Another innovation was the introduction of the lorry control park where lorries could wait their turn to be called to the sheds. This saved congestion on the quays and enabled the drivers to be told when they would be called.

By January 1966 the replacement of the old two-storey brick warehouses with the high roomy and modern transit sheds, together with the fine streamlining of cargo-handling ensured that the India and Millwall docks were among the finest in the world. Much was to change in the next four years.

92

RESEARCH INTO DREDGING THE THAMES

In 1949 Sir Claude Inglis with a team from the Hydraulic Research Station was asked by the Port of London Authority to review the problems of keeping the dredged channels clear. Between 1928 and 1956 the PLA had dredged 47 million cubic yards from the docks and river. By 1956 the cost was approximately £500 000 per year with no sign of dropping. Theories had been expressed that the practice of dumping dredgings in Black Deep, some 30 miles to the seaward of Southend,

Mowlem

Glengall Grove Bridge

might have had some connection with the amount of silt deposited in the upper reaches.

Following the building of a small pilot model in a transit shed on the south side of Victoria Dock to provide data on tides and currents, a much larger model was constructed in 1950–51. It was to a scale of 1:600 and nearly 400 ft long and 44 ft wide at the seaward end. Model tests were carried out, and later confirmed by full scale

93

tests in the river itself using radioactive isotopes encapsulated in finely ground glass particles.

Salt water is heavier than fresh water and the tests established that when the tide makes the fresh water flows over the top, and the salt water, heavily silt-laden, flows underneath, depositing suspended mud in the central part of the estuary, particularly in the Gallions, Barking and Halfway Reaches. At the landward and seaward ends the water was comparatively clear. It was also noted that there was a higher concentration of solids in suspension on the ebb tide than on the flood tide in the central portion of the estuary. Much of the material dredged to improve the channel for shipping had been dumped in Black Deep with the idea that it would be carried out to sea. The investigations showed that it was swept back up the estuary near the bed, being deposited whenever conditions were favourable. As a consequence, the more dredging done at Gravesend, the more mud was carried up river to Barking Reach, necessitating dredging on a large scale to maintain the 27 ft deep shipping channel.

Aerial view of Fred Olsen development, Millwall Dock

PLA

Unloading tomatoes, Fred Olsen terminal, Millwall Dock

The recommendations of the study were simple: dumping in Black Deep should cease and all future dredgings should be pumped ashore. It was thought that if this was done, the regime of the river would gradually settle down naturally and little further maintenance dredging would be required.

The research had been carried out over a period of eight years, but the results were far-reaching and conclusive. Even by 1952 preliminary results had shown that in future all dredgings would have to be pumped ashore, and the PLA acted immediately. The site chosen for receiving the dredging was at Rainham in Essex where the PLA owned 260 acres of land. As it was in the centre of the dredging area a considerable saving would be made over the cost of transporting to Black Deep. The ground comprised soft estuarine clay and peat overlying sand and gravel, and after the bund [embankment] walls had been built it was necessary to load the area slowly to balance the porewater pressure in the clay and allow settlement to take place. The pumping plant was situated at the middle of the river frontage on a pontoon moored between dolphins. A light jetty carried a 27 in. pumping main to the shore. The disposal pumps were of 1500 tons per hour capacity pumping through a distance of 3500 ft. It was originally thought that the area available would be full up in 7½ years but 20 years later there is still ample capacity to accept more.

TILBURY THE CONTAINER PORT

John Lunch, when Director General of the Port of London Authority, had pointed out that the policy should be to find out what the customer wants, and then provide it ahead of the actual demand for the service by a shipper or merchant. He also pointed out that port facilities are long-term assets, and should be filled with traffic at the earliest possible moment.

These two aims, admirable as they are, did not take into account the rapid changes in the public's demand for new and faster methods of travel.

Tilbury dock had for many years been the United Kingdom base for the P & O and the Orient Lines, and both had built new ships of 29 754 tons gross weight. Those of the P & O line were *Himalaya, Iberia* and *Arcadia*, those of the Orient Lines *Orcades, Oronsay* and *Orsova*, all constructed after 1945. By tradition the P & O occupied the south side of the main dock, and the Orient Line the Centre Branch Dock. The main dock at Tilbury is 600 ft wide and as it would not be practical for the new vessels to manoeuvre into the branch dock, it became evident that a new ocean terminal would be required to accommodate the liners. The main dock could be widened to 900 ft by demolishing the old quay wall of the West Branch Dock, which would enable ships to be turned and berthed on its east side, and in this way two additional berths would be obtained.

The contract was let to Holloway Brothers Limited. Work commenced on 8 June 1953 and was completed by 30 June 1957. Referring to the new venture, the Chairman of the PLA said at the opening 'In the design of this extension we made the best estimate we could of the developing requirements of the great ships which now use this dock. It must necessarily rest with others to take advantage of the facilities provided. We have performed this act of faith, and we hope for the helpful understanding and collaboration of all concerned', but in the four years that it took to build the new terminal the Comet aeroplane had gone into service and introduced the travelling public to the jet age. They could now no longer find the time to travel to Australia by ocean liner, spending three to four weeks at sea, when they could get there by jet in 24 hours. The days of leisure travel were over. Passenger liners no

Artist's impression of new Tilbury development

longer came to Tilbury and the few remaining in service transferred to holiday cruising based at Southampton.

In reviewing the development of Tilbury as a container port, the construction of No. 1 Berth is important, however, as it formed the basis for the design of the new branch dock.

The best site for a major development seemed to be the vacant ground north of the main dock where a new branch dock wide enough to take the largest vessels which could use the entrance lock could be built. The width would be about 600 ft, leaving a tongue of land 800–900 ft wide between the West Branch Dock and the new dock. The problem lay in how to deal with the southern extremity of this land. Between the West Branch Dock and the inner knuckle of the lock, the main dock

97

had been dredged to a 1:5 slope, because of the 50 ft depth of soft clay and peat. In order to turn ships possibly 800 ft long into the proposed new dock, the main dock would have to be widened to 900 ft.

The PLA designers considered 16 different methods of quay construction, including open-decked piled walls, cylinders and monoliths. Eventually it was decided that the most economical scheme was the construction of a quay wall 842 ft long using monoliths. These had previously proved successful, rapid and economical at Tilbury. An important consideration was the small amount of reinforcing steel required at a time of national steel shortage. The quay wall was turned at the west end for a distance of 240 ft to form the start of the new branch dock and thirty monoliths 30 ft square with four 10 ft 6 in. wells and chamfered corners were used. The monoliths were composed of mass concrete, 'allin' pit ballast being used for the aggregate. The bottom sections had reinforced shoes with 9 in. wide cutting edges and the bottom seals were 5 ft deep in the back and 12 ft deep in the front wells. They were all founded a minimum of 2 ft into the gravel, 12 ft below the ultimate dredged depth of 42 ft 6 in. below THW.

Monoliths 12 to 18 were founded in the chalk where the gravel layer is thin, and where a gulley crosses the dock from southeast to northwest. This has already caused problems in the building of the original dock, and in the first dock extensions at 31 to 33 berths, and was to cause trouble in later developments.

The Contractors had provided 2000 tons of 6 ton cast-iron kentledge blocks, shaped to suit the 3 ft width of the monolith walls and keyed together when stacked. A total of 33 blocks weighing 205 tons could be placed on a layer. The monoliths were sunk alternately to ensure equal conditions on opposite faces during the initial stages of sinking. The space between the monoliths was 4 ft 4 in. wide and closed at the back by four 18 in. octagonal concrete piles driven to the gravel in arch form, bearing on the corner chamfers of the monoliths. The average rate of sinking was 33 ft 4 in./week or 2.72 in./hour and the average skin friction during sinking was 8.18 cwt/sq. ft. No difficulty was experienced in excavating through the peat, and the rate of sinking was 50% faster than that achieved in 1913.

PROPOSALS FOR FURTHER DOCK EXTENSIONS AT TILBURY

The proposals for a mile long West Dock at Tilbury had first been formulated by Sir Frederick Palmer in his report of 31 December 1910. In 1958 these proposals were examined again. The first stage provided for four berths and the construction of a new impounding station. The impounding station was to have sufficient capacity to deal with the larger dock area, and to be capable of further raising the impounding water level by 2 ft 6 in. It was also to be designed and situated so that the minimum of silt would be pumped into the dock during impounding. The PLA's revised development plan of 1964 also envisaged a new quay nearly one mile long, at an estimated cost of £20 million, with four new berths to be built initially, and a new road system within the docks to link with the Tilbury Docks approach road and the new Dartford-Purfleet tunnel due for completion in 1963. It was

Congestion in Albert Dock in 1962

intended that the new dock would draw traffic away from the older docks closer to London and thus help to relieve congestion there.

Work started in 1963, and by the spring of 1965 the construction programme for the first stage of the new branch dock was well advanced, and the four berths had been allocated. The 1965 Port of London Plan recognized the need for a further extension to the new branch dock, and in May approval was given to proceed with stage 2, the construction of six additional deep water berths. In February 1967 during the construction of this stage, it was decided to go ahead with stage 3, taking the new dock up to the northern extremity of the PLA's land. This would provide two further 850 ft berths and a total length for the new dock of 4773 ft. These decisions, important for the future pattern of London trade, were taken before the long-awaited National Ports Council's *Interim Plan for Port Development* was published in the summer of 1965. In their report the NPC stressed the importance of reconstructing the major London export berths. By doing this, road vehicles could be dealt with more expeditiously, and cargo could be handled by mechanized methods more cheaply. Endorsing the initiative that the PLA had taken in rebuilding the sheds in Millwall Dock, the NPC thought that the first priority should be the redevelopment of the north side of Albert Dock. A dock of truly modern layout required a great deal of land (15–20 acres per berth), but as the PLA owned a substantial area to the north of Albert Dock, it was thought essential that this land should be brought into use for port purposes as soon as possible.

The shortage of capacity which had caused congestion in the docks could be overcome by the provision of 10 remodelled berths as was suggested for the North Albert project. This would also provide for perhaps two million tons of more of general cargo in the next decade. However, the NPC recognized the need for additional deep water accommodation for bulk cargo carriers. They approved the steps already taken by the PLA in going ahead with stage 2 of the new branch dock at Tilbury, but considered it to be too limited. Steps should be taken to provide fourteen new berths as quickly as possible by joining up the river cargo jetty to the Passenger Landing Stage.

As development of Tilbury Dock kept the PLA Engineering department fully engaged, the North Albert scheme was given limited consideration and it was allowed to be dropped from future plans. As things turned out this was a sound decision.

CONSTRUCTION OF NEW TILBURY BRANCH DOCK

In April 1963 it was decided to invite tenders for construction of quay walls for two new berths to the north of No. 1 Berth, totalling 1373 ft, and for the dredging of the first stage of the branch dock. The material dredged out for the new dock would be dumped in the Black Deep, as although all maintenance dredging was pumped ashore there was no capacity to take contract dredging. It was thought that the material, being solid, would not be so readily transportable by tidal currents. Later some material was pumped ashore at Cliffe, about six miles downstream from Tilbury. A new site investigation had shown that, in general, conditions were no worse than those encountered at No. 1 Berth and monolith construction was to be used as before.

On the previous occasion the rate of monolith sinking had been restricted by the need to load the units with up to 1150 tons of kentledge blocks to overcome skin friction. The kentledge naturally had to be removed before further lifts of concrete could be cast. A tender submitted by John Howard & Company Limited provided for monolith sinking using bentonite lubrication. This had only recently been introduced into civil engineering construction and was considered to be a daring innovation. The tender submitted by John Mowlem & Company Limited relied on large precast interlocking monolith sections, each weighing 20 tons. This would save time in monolith sinking, but would require a much larger amount of kentledge that the John Howard scheme, so the contract was awarded to Howard.

The monoliths were similar in design to those used in No. 1 Berth. Sinking was started by breaking out the concrete base upon which the shoe and first lift had been cast, and continued by grabbing out material from the four wells, the amount excavated being controlled to ensure even sinking. When the second 9 ft lift had been concreted, the monolith was sunk a further 8 ft leaving about 4 ft above ground level to avoid the risk of oversinking. The bentonite was pumped through twelve ¾ in. dia. pipes, cast three to each wall, with outlets above a 2 in. wide

100

ledge formed in the shoe. Pumping commenced after each stage of sinking and was continued until the bentonite could be seen at ground level.

The introduction of bentonite considerably reduced the skin friction as monoliths sank under their own weight even for a limited distance into the chalk. The rate of sinking was 47 ft/week for stage 1 and 64 ft/week for each side of the dock for stages 2 and 3, almost twice the rate achieved by Holloways in 1957. Although the PLA required a test load of 250 tons after the sinking of each monolith, a saving of half a million crane lifts had been made by using bentonite.

BERTHS 35–38

The two sheds on 36 and 38 berths were each 550 ft long by 150 ft wide, with a 50 ft wide canopy at the back. The overall design adopted for the sheds incorporated a monopitch roof on cantilever columns. The advantages of this type of construction for the quayside shed were:

(*a*) simplicity of fabrication and erection;

(*b*) only one line of roof gutters was needed with short outlets to the dock; and

(*c*) the area of roof and gable sheeting was reduced to a minimum.

The required headroom of 31 ft 6 in. for mobile crane working under the rear canopy could easily be provided with a roof slope of 1:19. Each shed would take 3000 tons of export cargo.

Monolith sinking for stage 2, New Branch Dock, Tilbury

PLA

At the end of 1962, discussions were held with the Atlantic Steam Navigation Company for the provision of two roll-on/roll-off berths in the new branch docks. These were to be purpose-built berths to suit new vessels ordered to take through lorry traffic to the continent, and were to be situated on the west side of the new branch dock opposite sheds 36 and 38. As the ferry vessels were of shallow draft, it was not considered necessary to construct a monolith wall, but to slope the excavation at 1:5 down to the dock bottom under the jetty. However, this slope was 1000 ft long and 25 ft wide and proved to be unstable in the dry. After a slip had taken place it was trimmed to 1:7, the toe was weighted with concrete hardcore and the surface was covered in asphalt.

A false quay of a flat slab on piles offered the simplest and most economical form of construction: 20 in. diameter BSP cased piles were used for the jetty, driven vertically, with rakers at bollard positions. The piles varied in length from 58 ft to 114 ft where a thin gravel layer was met near the middle of the jetty head. This type of pile could easily be extended by welding on additional lengths of pile casing, and redriving.

By this time the old bund wall had been breached and the dock area flooded to enable dredgers to work. This meant that all piles for the jetty had to be driven from floating craft. Problems also arose with the striking of the soffit shuttering to the deck when the dock water was at impounded level. Usually jetties are built in tidal waters and the carpenters are able to strike the shuttering at low water. In this instance there was only three feet between impounded water level and the soffit of the deck, and a system had to be developed whereby shuttering could be released from above. This was achieved by concreting yokes with crossheads into the tops of the piles from which the shuttering was hung. When the deck concrete had hardened, the suspension bolts were released allowing the shutter panels to be withdrawn using a crane standing on the completed deck. By careful planning of the concreting sequence it was possible to remove all the panels fairly easily. The contract was let to John Mowlem & Company Limited on 28 January 1965 and completed on 30 June 1966.

RECLAMATION FOR BERTH AREAS

Monolith sinking for stage 2 had begun in August 1965, and by November sufficient progress had been made for John Mowlem & Company Limited to start on the reclamation contract for new berths. This included bunds, filling, roads, railways, drainage, electrical services and the diversion of surface and foul sewers.

Most of the ground to the east of the new dock was marshland with a number of ponds with a general level between 1.5 and 3.0 OD. At the head of stage 2 of the dock extension, where the Stewart and Lloyd assembly sheds had been built for the wartime PLUTO scheme, the ground was at 8.0 OD. To the west of the new dock there were ash hills 35 ft high above the surrounding marshland, where London ash, clinker and rubbish had been dumped until about 60 years before. This material was levelled down to the new general level of + 18.0 OD and provided much of the filling to the west side of the dock.

The Botney sluice was the main surface water drainage ditch for the whole of the marsh area between Tilbury and the higher ground at Chadwell St Mary. It crossed the whole site but had to be kept open at all times. Diversions were made and finally it was provided with a new outfall in North Tilbury.

The only means of access to the berths on the south side of Tilbury main dock was by road over the bascule bridge at the main entrance lock. Before this road could be removed a new temporary road and railway had to be laid on a 14 ft high embankment along the eastern boundary of the PLA estate, crossing the head of stage 2 and following the west side to join the existing road and railway at the bascule bridge. Time was of considerable importance, as dredging could not begin until the new

Mowlem

Marsh land at Tilbury, before reclamation

access had been established. The marshy ground to the east provided the most difficult problems. Temporary roads had to be made across the marsh so that lorries bringing hoggin and Thanet sand could reach their destination. Large pieces of masonry from the gate platforms on the Western Entrance to Victoria Dock which was being rebuilt at the time were used to fill the ponds. As there were 33 ft of soft clay and peat overlying the gravel, considerable settlement was to be expected, and in fact settlement plates built into the embankment recorded 3 ft 3 in. after eight months during the dock construction period. Stelcon raft slabs were laid as a temporary roadway surface, and stood up well to heavy traffic with a limited amount of repacking. The 24 ft wide road (including the laying of a rail connection to take boat trains to 33 shed) was completed by 30 June 1966, a period of 34 weeks.

Before reclamation could begin, bund walls had to be built to contain the surplus water. The main eastern bund was provided by the access road embankment, covered with Visqueen plastic sheeting to ensure that no water penetrated under the roadway. Pump water drained to sumps and was pumped back to the river.

A sub-contract was let to the London Dredging Company to provide over 2¼ million cubic yards of material to be pumped ashore over an area of 150 acres. Thames sand and gravel were taken from Lower Hope Reach six miles down river from Tilbury where the channel was being deepened as part of the Authority's dredging programme. This was dredged by a trailing suction dredger, deposited on the river bed near the new works and pumped ashore by a cutter suction dredger through a 24 in. pipeline. A maximum rate of 56 000 cu. yd/week was achieved, the pumping distance to No. 40 berth being 4600 ft. As the material was of excellent quality for concrete aggregate a stockpile was built up and used for nearly all the concrete in the monoliths, paving and reinforced concrete work. The angular shape of the flint aggregate caused considerable wear in the pipelines, however, so in later stages sand from further down the estuary was used.

NEW IMPOUNDING STATION

The capacity of the old 1929 impounding station was inadequate for the requirements of the considerably enlarged dock, and the intake in the tidal basin was badly situated so that large quantities of silt were drawn into the dock. When the installation of a new impounding station was authorized, tests were carried out to find the best situation. In the area of Northfleet Hope upstream of the main entrance lock clear water was found, except at bed level, for a period of about five hours, and special intake heads were designed to collect this for a period each side of high water.

It was the PLA Engineer's intention that a circular pump chamber 100 ft deep should be sunk well into the chalk to provide adequate cover for two separate tunnels to be driven under compressed air, and tenders were called for on that basis. John Howard & Company Limited submitted an alternative which was considerably cheaper than the Engineer's scheme. The 47 ft dia. pumphouse would be sunk using bentonite to a depth of 75 ft only. The two intake tunnels would be re-designed and placed vertically one over the other, so that they could be constructed in a steel sheet piled cofferdam. The excavated level of −49.0 OD ensured that the culverts were founded in the gravel.

The pumphouse was sunk without difficulty using bentonite. Although the ratio of weight to surface area was considerably less than the monoliths, no kentledge was necessary and it ended up only three inches out of plumb. Considerable pumping was necessary to 'bottom up', four 6 in. pumps being used with a network of drains. If the original scheme had been used it is likely that the bottom

PLA

Reclamation for New Branch Dock, Tilbury

would have been concreted under compressed air. For the intake six cofferdams were needed, each about 120 ft long, and 80 ft long piles were used for dam No. 1 with about 5 ft penetration into the chalk. It was unfortunate that this cofferdam crossed the gulley in the gravel which has caused so much trouble. There was little toe resistance, and the bottom of the dam blew up, serious blows occurring where the piling abutted the pumphouse monolith and at the northwest corner closure piles. The works were considerably delayed and fifteen months elapsed between the start of sheet piling and the eventual dewatering of the cofferdam. Finally this cofferdam was divided transversely for the remaining 7 ft of excavation, and the bottom concrete was placed under water. In the remaining cofferdams 90 ft long piles were used, and the work was completed without difficulty. In the ten years that the new impounding station has been in operation, no major maintenance dredging has been required in the dock.

BERTHS FOR FOREST PRODUCTS

During 1976 packaged timber in unit loads was being adopted worldwide. The PLA decided that if London was to remain the leading UK port for handling forest products, some berths in the new branch would have to be allocated to the timber trade, and long before stage 2 of the dock extension had been completed, two berths were allocated to handling bulk timber cargoes.

Number 42 berth of 20 acres was taken on a long lease by MacMillan Bloedel of Vancouver. The Consulting Engineers, Philip Reid and Walpole, decided that the best way to pave such a large area was to use 6 in. double-reinforced concrete, divided into 24 ft square bays with dowel bar joints between adjoining bays. It was thought that this would provide maximum flexibility to accommodate the settlement that was likely to take place under heavy loading. Since a high output had to be achieved, the contractors, John Mowlem & Company Limited, decided to use a 'paving train' with the dowel bars positioned in the road forms. The transverse

Dredging for New Branch Dock, Tilbury   *PLA*

joints were pre-assembled with the dowel bars set so that the paving train could pass over them. While the concrete was still wet the top of the joint was positioned. At the end of each run the paving train was picked up, using an excavator with a long jib, and set on its next track, ready to make the return run.

In September 1967 the first ship to enter the new section of the dock was the *Laconia* with 18 000 tons of packaged timber, closely followed by the *Vinni* with 22 000 tons.

Number 44 berth was let to the Svenska Cullulosa Company of Sweden. This berth of 22 acres required a large covered area for storing paper and wood pulp. John Laing were the Contractors and for this berth dry lean concrete with 'black top' surfacing was chosen.

Number 46 berth was let to the Seaboard Pioneer Terminal, a member of the Reed Paper Group. The berth, originally 17.6 acres, was later extended to take in 46A

106

berth, an additional 12 acres. The main area was originally marshland and ponds, so considerable settlement was to be expected. The PLA specification of 10 in. of dry-lean concrete with 2½ in. of hot rolled asphalt base course and 1½ in. wearing course of hardened asphalt to 'bus-stop' specification was used successfully.

In the older dock area, No. 34 berth was redesigned to permit the handling of packaged timber. An area of nine acres was cleared to the south of the main entrance lock and laid to concrete. This area had remained undisturbed since McAlpine's had built the western entrance lock in 1928, and further settlement was not anticipated. Soft spots were filled with well-rolled hardcore, and a 6 in. thick reinforced concrete slab was laid using interlocking joints between the bays. In 1977 it was decided to streamline the berth by demolishing the front halves of sheds 34A and B to provide a quay apron 104 ft wide. In 1978 over 200 000 tons of forest products were handled in package form by this berth.

The success of modern timber-handling methods can be seen by the fact that in 1976 the four Tilbury berths handled 1.5 million tons, compared with a total of 0.8 million tons for the whole 36 berths in the old Surrey Dock.

THE GRAIN TERMINAL

Just as bulk carriers had been built for the forest products, changes were taking place in the transport and handling of grain. The new bulk grain carriers of up to 35 000 dwt (deadweight tonnage) were far too large to go to London's upper docks. The PLA for many years had been proud of the floating grain elevators and the Millwall Dock Central Granary, but all these would become redundant. The only way of coping with changing conditions was to forget the past and concentrate the grain-handling business in one terminal on the river front at Tilbury. Here it would be possible to have a minimum water depth of 42 ft which would accommodate 35 000 dwt carriers. Later if bigger bulk grain carriers up to 45 000 dwt were to become common, they could be accommodated by increasing the dredged depth.

The site on the river side of No. 45 berth was chosen as there was room to extend the terminal, and tenants could set up their own flour mills nearby. Extensive investigations into the economic, financial and technical aspects of grain handling were carried out. Meetings with trade interests and others concerned were also held and as a result it was decided that the proposed 900 ft long jetty should house ship discharge equipment which would operate with any draft of ship at any state of the tide. The main feature would be two combination dual-discharge towers 180 ft high, each with a bucket elevator with an unloading rate of 1000 tons/hour. This rate is considerably higher than at any other terminal in Europe, and was based on successful Canadian experience. The fast discharge rate was to be matched by a conveyor system that could deliver grain to coastal vessels, barges, road or rail vehicles or to silo storage. Nine lorry loading bays and three rail loading bays could handle up to 1200 tons/hour. A central control panel with pushbuttons to direct the conveyors was to be installed with mimic flow panels to provide a continual pictorial check on actual flow paths. The original silo structure ashore was to consist of one bank with a storage capacity of 35 000 tons. However, because of marketing considerations the capacity was later increased to three banks, with a capability for further expansion if necessary. Each bank is 256 ft long, 80 ft wide and 220 ft high. Two of the bins are subdivided to provide a variety of bin sizes from 60 tons to 1000 tons, but the third has a preponderance of larger divisions only.

Shoesmith Howe and Partners, who had experience of similar installations at the Great Lakes in Canada, were engaged as consultants for the design of the complex. The contract was awarded to John Howard & Company Limited.

Berth 34, packaged timber, Tilbury

As the loads from the marine unloading towers were high, the consultants decided to use cribs sunk on to a prepared section of the river bed. This method had worked well in the still waters of the Great Lakes. The cribs were of cellular construction and were constructed in the larger of the two old dry docks using slipform methods, a layer of broken stone being laid on the dock floor to ensure flotation. As the units were 60 ft high it was necessary to concrete them up to half height, float out, and complete them afloat. When the first one was towed out and sunk on to its prepared base such was the intensity of scour that it tipped over at an alarming angle, and the scour problem was only overcome by driving sheet steel piles into the river bed around the cribs.

108

The extensive mechanical and electrical equipment for the silos took six months to commission, but when everything had been sorted out and adjusted the terminal was an unqualified success after it was opened in June 1969. Adjacent to the silos, four privately owned mills have been built on land leased from the PLA. These are owned by Spillers French, Rank Hovis and Allied Mills and all draw their grain direct by means of conveyor belt. Additionally, a starch products mill owned by Cargill-Albion which is also served directly from the PLA terminal, has been built.

CONTAINER BERTHS

Containers were devised many years ago by Pickfords for furniture removals. The firm even had arrangements with the railway companies and shipping lines for furniture to be sent overseas in their private containers.

The containers could not be universally adopted until standard external dimensions, ratings, specification, testing procedure, marking and corner fixings had been decided. This was essential for quick and easy transfer between all forms of surface transport. The accepted sizes were 8 ft wide and 8 ft high, with lengths of 10, 20, 30 or 40 ft. Once these had been agreed internationally at a Conference in 1965, the scene was set for the Container Revolution.

With containerization in mind, the PLA was able to develop the berths on the west side of the new branch dock at Tilbury relatively cheaply for this method of cargo handling. Investigations into Port Planning showed that a conventional berth with cargo stacked on pallets would have a potential throughput of ¼ million tons using between 6 and 7 acres. A container berth with 850 feet of quay and about 20 acres of back-up land behind could handle about 1 million tons/year. Each container berth required a considerable investment in equipment. A Paceco Container crane costing £¼ million then would be required for each berth, together with straddle carriers and trailers for transferring those containers earmarked for transport via the Freightliner Terminal which was to be built by the PLA and British Rail on the east side of the new branch dock.

John Mowlem & Company Limited won the contracts for the development of the container berths, starting with berths 40 and 43. For all the berths the best way of laying large areas of dry lean concrete were explored. Initially a Blaw-Knox PF 90 paver was used for spreading, laying and tamping the concrete. It worked satisfactorily on berth 43 where the formation was mainly rolled ash, but on berth 40 where the surface of the pumped fill was loose sand and gravel, the machine tracks dug into the formation, leaving unacceptable ridges on the surface. The best way to lay large areas of dry lean concrete was found to be by using long 12 in. by 8 in. timbers as guides or screeds. The concrete arrived in end-tipping lorries and was levelled by tractor shovels. Compaction was by vibrating roller, and timbers were towed from section to section. Outputs in excess of 3000 cu. yd/week were obtained using very few men. The rolled asphalt paving was laid fairly close behind the dry-lean concrete sub-base to prevent damage. A hardened asphalt wearing coat was used

PLA

Aerial view of Tilbury Grain Terminal

to prevent the pads on the bottom of the containers from penetrating the surface when stacked two high.

The outer rail for the container cranes was laid on the edge of the monolith quay. The back rail was laid on a special beam supported on 22 in. dia. cased piles driven into the sand and gravel layer above the chalk. Secondary beams spanned between the crane beam and the back wall of the monoliths. To help progress, a number of early beams were precast off the site, but later they were all cast in situ using block-work back shutters. The crane rails were 123 lb/yd flat bottom, continuously welded by the Thermit process. Initially the rails were laid on a cement grout bed, but later, to prevent local crushing of the concrete, an epoxy resin base was used.

Berth 39 was selected as the site for the UK sea terminal of Overseas Containers Limited (OCL) which was formed in 1965 by its four principals (British and Commonwealth, Furness Withy, Ocean Fleets and P & O) to develop the containerization of the UK – Australian shipping trade. By January 1967 they had decided to build six container ships each with the carrying capacity of 1300 containers.

Following the decision by OCL, Associated Container Transport was formed by Ben Line, Blue Star Line, Cunard and Ellerman Line and agreed in 1968 to build three ACT container ships. They arranged to work from berth 45 at the head of the dock. Two groups agreed to run a joint service, each carrying the other's containers, thus establishing a weekly container service between UK/Europe and Australia/New Zealand. After delays caused by labour problems the service was commenced on 18 May 1970 with the *Jervis Bay* loading containers at Tilbury.

OCL engaged Sir Bruce White Wolfe Barry and Partners as Consulting Engineers to develop 39 berth and to look into the problems of stacking containers five high, and also to advise on the provision of centralized refrigeration for full meat containers waiting at the berth. Much of the original thinking in the pioneering work done by OCL has been now superseded by later methods and plant, but the original plan was for cargo to be collected and packed into containers at a packing depot at Orsett just outside Tilbury, where items could be handled under cover. Full containers would then be transferred to the Tilbury Terminal at 39 berth and stacked five high using electric overhead travelling cranes, the stack being an inversion of the ship's stowage plan.

Calculations confirmed by wind tunnel tests showed that empty containers stacked more than three high could be blown over. This meant that protection would have to be given if containers were stacked higher. For perishable goods special insulated containers were designed which permitted chilled air to be circulated through the container to maintain its low temperature. The refrigerated stack consisted of a fabricated cell structure similar to that used in a ship's hold.

As an overhead gantry crane was to be used for stacking the containers it was decided that cladding for wind protection could best be provided by using the crane support structure. This was cheaper than the provision of guides, which were considered to be unnecessary except in the refrigeration area where accurate location of the cold air ducts was essential.

As a stack of containers 40 ft high would be sensitive to settlement, the whole of the stacking area was piled, one pile under each corner of the container pads.

The quay was 850 ft long with an apron 140 ft wide. This area is served by one ASEA twin lift and one Paceco twin lift crane. Both are 45 ton capacity and capable of lifting two 20-ft or one 40-ft container at a time. The cranes have an outreach of 110 ft, and a back reach of 50 ft.

The stack shed behind the quay was 860 ft long and 170 ft wide, with a capacity for 2000 20-ft containers stacked five high. The stacking area had been divided into four blocks with a storage capacity for 360 refrigerated containers in the southwest corner, the cool air being provided by a plant similar to that fitted in the ships. The refrigeration was effected by three reciprocating compressors, two 170 hp and one 250 hp.

Behind the stack shed there were four acres with a storage capacity for 480 containers stowed two high. However in a very short time the whole method of container handling changed. When ACT took over 45 berth for their terminal they decided that containers should be handled by straddle carriers, and stacked no more than two high. Instead of the elaborate refrigeration installation, much smaller HOLIMA units were used. These comprised a central refrigeration unit, with two arms taking two tiers of six containers each side, making a total of 48 in all.

111

PLA

Northfleet Hope development, Tilbury

NORTHFLEET HOPE DEVELOPMENT

The success of the OCL and ACT ventures was such that by 1976 they were looking for further expansion. A joint investment between the PLA, Overseas Containers Limited and Associated Container Transportation (Australia) Limited was formed to reclaim 25 acres from the river at Northfleet Hope. This would provide a deepwater berth with a quay 1000 ft long and 42 ft minimum depth of water, sufficient to take the next generation of container ships. The complex, costing some £24 million including equipment, was planned to take 75% of the total liner trade between the UK and Australia and New Zealand. With the redevelopment of 37 berth, and the taking in of land to the west of it, the terminal covered 64 acres, all of which was paved. This provided a vast container park with a capacity of 6104 containers. The area of 37 berth was devoted to unit refrigeration plants, similar to 45 berth. This would make the complex the largest refrigerated container stack in the world, accommodating 1464 units. When fully operational it was expected that the new terminal would have an annual throughput of about 170 000 containers. This was a major factor in maintaining the position of Tilbury and the Port of London as the premier container port in the UK, with the capacity for handling the largest container ships afloat. The terminal would be operated by Tilbury Container Services, a joint operating company formed by OCL and ACT.

The design prepared by the PLA Engineering Department envisaged the reclamation of 25 acres from the river bed in an area between the entrance lock and the Grain Terminal known as Northfleet Hope. A dredging contract was let to Sir Alfred McAlpine Zanen to dredge the berth and to obtain gravel filling from nearby reaches and dump it on the foreshore. Over one million cubic metres were obtained in this way.

The filling, some 480 ft in front of the old river wall, was retained by means of a wall of Unissen piles. These were heavy universal beams with clutches welded to the front flange so that they could be interlocked together into a continuous line. They were tied back to a row of anchor piles buried in the filling with a tie rod connected to every pile at low water level and to alternate piles at coping level. The lower tie rods were embedded in a prestressed concrete waling formed by concreting between the webs of the piles. Stressing cables were threaded through holes cut in the piles and stressed in long lengths. Although the designer thought this to be an ideal solution to his problems, practical difficulties of carrying out high quality 'prestressed' concrete work at the lowest of spring tides, with only two hour's occupation per tide, were encountered but overcome.

The suspended quay, 100 ft wide in front of the Unissen piling, was of the heaviest construction because two second generation container cranes would have to run along it in addition to straddle carriers: 30 in. dia. steel tubular piles each carrying 200 tons driven with flat shoes up to 30 ft into the chalk were chosen. This required piles up to 130 ft long. Two rows carried the heavy coping beam and outer crane rail, and two rows of raking piles carried a heavy intermediate beam taking the outer end of the quay and also forming the seating for Ministry of Transport standard bridge beams spanning to the Unissen piles. The fender piles were H section with 2 in. flanges over 100 ft long, housed in recesses in the quay coping and fixed to Andre fender rubbers.

As the ground on the landward side of the reclaimed area in the neighbourhood of the old river wall had never been subjected to superimposed loads it was decided that it should be loaded with ¼ million cubic yards of Thanet sand piled up to 8 ft high. This would induce rapid settlement if left in position for a year, and so limit maintenance problems later when the berth was in operation. After it had served its purpose it would be spread and levelled to make up the formation in the fill area.

As the proposed works had received considerable publicity as many as eighteen contractors asked to be included on the tender list. All were interviewed by the Chief Engineer to the PLA and a short list of seven major national contractors was drawn up. Tender documents were issued in May 1976. With strong competition for this contract, it was most important to consider the planning of the works, particularly as time was short, and it was unlikely that an extension would be granted.

The thinking behind the scheme of the successful contractor, Mowlem, may be of interest.

Berth 45, Tilbury, showing HOLIMA units

Much of the tender sum was in the supply of permanent materials, common to all firms tendering. It was essential therefore that the best prices should be obtained coupled to firm delivery dates.

As more than 85 acres of dry lean concrete had already been laid in Tilbury Dock, the fastest and most economical methods of laying were known. Since 65 acres of asphalt were required, it would be necessary to obtain subcontract prices from all firms capable of laying such a quantity in a limited period.

The key to the award of the contract appeared to be in the method of construction used for the false quay. Much of the work would be over water, so the amount of falsework required would have to be reduced to a minimum. Heavy lifting plant would be required for handling and driving the piles (a Delmag D 44 diesel hammer and pile together weighed 18 tons), so plant could be made available to handle large precast units (up to 20 tons each).

Soffit and edge beams would be precast off the site, but the positioning of all piles would have to be done accurately so that everything fitted together without trouble. To ensure accurate pile positioning, a special pile gantry travelling over the driven piles could be used for locating and driving a group of six piles, before moving to the next section.

A 25-ton electric derrick and two 20-ton derricks were available, together with travelling gear and the necessary waybeams and steelwork to make the piling gantry. An additional 15-ton derrick was available for driving the Unissen piles.

This thinking not only achieved the award of the contract but in practice worked extremely well. In spite of a number of early setbacks, including a rogue ship which demolished the first two month's offshore piling work, the completion date set in 1976 when the scheme was first decided upon was met without undue difficulty. The *Encounter Bay* tied up on 15 September 1978 to discharge its first load of containers.

THE WEST AFRICA TERMINAL

By 1972 several factors had affected the south side of Tilbury main dock. The overhang on the self-unloading timber ships using 42 and 44 berths was enough to cause ships to collide with the bascule bridge when entering the dock, and its removal would help the navigation of larger ships into the lock. The old entrance lock and tidal basin had already been closed to shipping and with fewer barges, and fewer but larger ships, it was found possible to work the whole of the dock on one entrance lock. Considerable changes were taking place in the nature of West African trade. Imports from Lagos had risen to about 3¼ million tons/year, whereas exports had dropped to around 380 000 tons. To cope with this changing trade, the Elder Dempster and Palm lines and their associates decided to introduce new multipurpose semi-container ships. The change in vessel design and cargo-handling procedure meant that if the increase in ship operating and particularly cargo-handling costs were to be contained, all activities would need to be gathered into one terminal. The P & O had vacated their Australian passenger liner berths at 31 to 33 sheds and the redevelopment of the whole area was a possibility. By building a dam across the old entrance lock a new road could be taken to the berths from

Mowlem

OCL container stack, berth 39, Tilbury

115

the east, allowing the bascule bridge to be removed. This would also give access to the highly successful No. 34 timber berth.

By leaving the old outer lock gates in position and building a wall of blocks on the apron like that at Gallions, it was possible to construct a dam by concreting between the two. When this had been done a roadway was quickly constructed over the top of the dam. With the demolition of the old P & O office and sheds, the last traces of passenger services to the Far East vanished. The site was completely cleared and Thanet sand filling imported to make up levels. A portion of the tidal basin was also reclaimed by tipping Thanet sand in such a way that the mud was displaced, and surcharging this to limit the settlement. Later the road was re-aligned over it. The entrance lock was also filled, as was the dry dock. The boundary of the site was taken up to the small dry dock which was turned into a barge berth. This allowed barges to be loaded or unloaded away from the main quay, thereby increasing efficiency.

By these means a quay 2700 ft long was obtained which provided four ship's berths equipped with eleven portal cranes from 10 to 15 ton capacity. The total area of the terminal was 39 acres, paved in the standard PLA dry lean concrete and asphalt. A section on the east side was treated with 'Salviacim', a special asphalt surface treatment impervious to oil and abrasion from heavy lift trucks handling logs in the area.

The site was divided into two berths for export, and two for import. The export section contained one of the largest sheds in Europe, 1178 ft long and 155 ft wide clear span, though to comply with fire regulations a fire resistant wall divided it into two. The import area had 90 000 sq. yd of open storage for logs and timber and two sheds each 280 ft long and 156 ft wide.

FUTURE DEVELOPMENT

As can be seen, a vast amount of construction work has been carried out at Tilbury to keep pace with the ever-changing needs of shipping companies. Although Harry Dobee could not have foreseen the good use made of the 450 acres of marshland he bought at agricultural rates in 1882, he made a wise decision to buy so much. It has taken nearly 100 years to develop it, and there would still be room for more riverside berths if the tidal basin were completely filled in, or Northfleet Hope was extended. The old branch docks could be redeveloped and made into modern berths by filling in the centre branch dock and the space made available for cargo storage. A piece of ground known as the Fort Land to the east of Tilbury Landing Stage is also available for future expansion.

Riverside docks

WOOLWICH AND DEPTFORD

Although this book records the history of the enclosed docks, some mention should be made of developments which have taken place in the tideway. The famous dockyard at Woolwich built the *Great Harry*, the first ship in the British Navy, which was launched by Henry VIII in 1514. Naval ships were built here until 1869. The ordnance activities developed alongside the dockyard, and the Royal Woolwich Arsenal became the largest munitions factory in the country.

The Deptford dockyard was also founded by Henry VIII and became the leading centre of naval shipbuilding in Europe. The yard, some 30 acres in extent, had two docks and three slipways upon which naval vessels were built. By 1742 the Admiralty had established the Victualling Yard close by. After a disastrous fire, new buildings were erected in 1780. The river wall and dock entrance were rebuilt by John Rennie in 1814, and as a result of Queen Victoria's visit in 1858 the yard was renamed the Royal Victoria Victualling Yard.

The dockyard continued building naval craft until 1869, and was closed at the same time as Woolwich; with the advent of iron warships of greater draught both yards were too remote from the heavy industrial areas for shipbuilding to be economical.

The Royal Victualling Yard became a food supply depot for the forces during the 1914–1918 war, and later served as the army reserve depot. In 1959 it was rebuilt but since then the site has been purchased by the Greater London Council for housing, ending London's many years of association with the Navy.

REGENT'S CANAL AND DOCK

In 1812 on the north bank of the Thames in Limehouse, the Regent Canal Company built a barge basin close to the Thames with the intention of connecting with the Grand Union Canal. This would enable export goods to be taken from the industrial Midlands to the Continent without trans-shipment. Objections to the building of a canal through the residential areas of north London were as strong as those 150 years later over motorway construction. One writer to the *Gentleman's*

PLA

Deptford Dock

Magazine suggested that the extremely varied weather experienced over the country would be adversely affected by the increase in moisture caused by extensive canal building.

It is doubtful whether the Regent's Canal would have been built at all but for the passing of the Poor Employment Act of 1817. This enabled poor people to be found labouring work on useful projects. The Canal Company immediately took advantage of this Act and started construction in the same year. The basin was enlarged to take sea-going vessels and completion of the whole project was achieved in 1820. At one time the British Waterways Board, who since nationalization have been responsible for the canal and dock, ran a through service from Birmingham to Hamburg using their own small coaster-type vessels.

Traffic from the River Lea was fed into the Regent's Canal Basin through the Limehouse cut, a canal joining the River Lea just below Bow Bridge to its own entrance lock to the Thames close by. When the entrance lock became unsafe in the early 1970s, it was filled in, and all traffic diverted into the Regent's Canal Basin. The Limehouse Cut is now mainly used for rubbish barges from northeast London Boroughs taking refuse to dumping grounds at Rainham in Essex.

A major overhaul of the Regent's Canal Basin entrance lock took place in 1917 when new gates were fitted. No further renewals were needed until new sill timbers were fitted in 1953 when fresh outer gates were installed. However, by 1964 it was necessary to renew the inner gates, pintles and sill timbers. As the lock had to be kept open for craft the renewals of the sill timbers had to be carried out inside a horizontal limpet dam allowing shipping to pass over it at high tide. The limpet consisted of two main sections, a vertical shaft to fit around the hollow quoin, and a tunnel section leading from the bottom of the shaft along the sill to a distance five feet beyond the mitre point. All sections were reversible so that they could be used on either side of the lock.

POPLAR DOCK

Just as the Regent Canal Company established an outlet to the Thames at Limehouse, in the mid-19th century the Great Eastern Railway Company decided that it needed a similar outlet.

About four acres were rented from the West India Dock Company, and two small branch docks constructed in an area close to the Blackwall Basin. Access to the Thames was through the Blackwall entrance lock. Since the closure of Blackwall lock it had been necessary for craft to take the longer route through the main South West India Dock Entrance. The dock, now run by British Rail, is served by a rail link from Bromley By Bow and is still used as an interchange between rail and water. The lease inherited by the PLA from the West India Dock Company was for 999 years. Should the PLA wish to close West India Dock, providing access to this dock might be a problem.

Mowlem

Regent's Canal Entrance Lock, horizontal limpet

119

RESPONSIBILITY OF PLA FOR NEW CONSTRUCTION

Since the formation of the Port of London Authority in 1909, it has had the responsibility for approval of the plans for all new structures to be built in the tideway. The current Act states 'Under the Port of London Act 1968 Section 66–72 a licence is required from the PLA before works of any nature whatever are carried out in, under, or over the Thames or before the banks of the Thames are cut in any way'.

In this way the Authority has been able to keep a close check on all structures built on the foreshore particularly ensuring that there is no interference with the free flow of the tideway or navigation by shipping.

Other consents required may include for example, planning approval; river authority or GLC approval where floodwalls are affected; that of the Ministry of Agriculture, Fisheries and Food under the provisions of the Dumping at Sea Act 1974; and the approval of other riparian owners if their private rights are affected by the river works.

RIVERSIDE JETTIES

Before the 20th century, the majority of jetties were built in timber. The method of construction was usually to use 12 in. by 12 in. pitch pine piles, 12 in. by 12 in.

Mowlem

Inside limpet, new sill

120

Admiralty type jetty, Blackwall entrance

crown timbers and 12 in. by 6 in. bracings with bolted fixings. An advantage of this method was that ships could berth heavily against them without damage. With continual berthing the Gravesend landing stage became so flexible that it would move about two feet when the Tilbury ferry boat berthed heavily against it.

One of the disadvantages of using timber for jetty construction, particularly in the lower reaches of the river, was the presence of marine borers. The Teredo worm, which attacks any softwood, was brought into the tideway on the hulls of wooden warships visiting Sheerness dockyard during the 19th century. The area of activity has varied considerably over the years. During the 1950s when the river was heavily polluted, it was almost eradicated. Now with the considerable improvement in the cleanliness of the water, its sphere of activity is increasing and has now again reached Shellhaven on the north bank of the river. The timbers now available which are capable of withstanding the marine borers activity are ekki, spotted gum and greenheart, all of which have been used successfully for piles in recent years.

The use of reinforced concrete for jetty construction was developed in the 1920s. Sir Cyril Kirkpatrick as Chief Engineer to the PLA had seen the construction of the cargo jetty at Tilbury, and later adopted a similar design as Consultant for Ford's jetty at Dagenham. Both of these were supported on cylinders. Two other jetties built in concrete before World War II were the Samuel Williams jetty (also at Dagenham) and the Purfleet deep water wharf about four miles downstream from Dagenham. After World War II two further jetties were built using prestressed con-

121

Number 4 jetty Coryton

crete and cylinders. The No. 4 jetty at Thameshaven was designed by L. G. Mouchel and Partners in 1950 and at a time of steel shortage made use of the newly introduced method of prestressing by the Freyssinet system. The jetty, in the form of the letter L, has an approach 400 ft long with a head 250 ft long. It is founded on 23 reinforced cylinders of 7 ft 6 in. internal diameter, sunk some 30 ft into the river bed. As there is 34 ft of water at low tide some cylinders had to be up to 84 ft long.

The approach beams were precast in eight-ton sections on the shore, and lifted and placed in position on the reinforced concrete cylinder caps. Twelve high-tensile steel prestressing cables were threaded through the units and post-tensioned by Freyssinet jacks. A unique fendering system was developed by Professor A. L. Baker using precast prestressed suspended gravity fenders. These are 48 ft long and weigh 50 tons each; they were precast on the jetty deck and lowered some 25 ft to their final level.

The jetty was designed for 28 000 ton tankers and has been in constant use for nearly 30 years. Some of the prestressed concrete approach beams have been

122

replaced by steel sections encased in concrete after ship damage. The Baker fenders are still in good working order.

The use of prestressed concrete in jetty construction was taken further when Sir William Halcrow and Partners designed the Erith Jetty for Wm Cory and Sons Limited. This was for discharging bulk commodities of all types by means of grabbing cranes. Vessels up to 14 000 tons displacement could be accommodated. The jetty was again L-shaped with an approach 530 ft long, and head 557 ft long. The decking was supported on prestressed concrete cylinders 75 ft long, assembled vertically on shore, floated into position by pontoons and sunk by grabbing and kentledge. The jetty structure consisted of rigid portal frames with precast prestressed beams and deck soffits. The fenders were steel box piles with rubbing pieces.

When he was at the Admiralty, Mr D. H. Little developed a method of jetty construction using box piles with a simple four-foot thick deck slab. Fendering was by means of independent greenheart piles housed into a recess in the deck slab. The first use of this method in London was for the reconstruction of the Blackwall lead-in jetty to West India Dock in 1951. Other designers have elaborated on this basic idea using Rendhex piles and more recently tubular steel piles.

L. G. Mouchel & Partners, who have designed many of the oil refinery jetties at Shellhaven, and Mobil Coryton on the north bank in the Thames estuary, have had to provide for the berthing of 200 000 ton tankers in their design. A depth of 70 ft of water at the jetty head meant that No. 6 Rendhex piles up to 140 ft long were required. With approach arms up to 1000 feet long, much use has been made of precast concrete pile caps and decking with the minimum of in-situ work.

The berthing forces have been taken by robust breasting dolphins set some five feet proud of the jetty head. Tubular piles 3 ft 9 in. in diameter and 110 ft long, weighing 23 tons each, were used in the rebuilding of No. 4 jetty Coryton. The most recent breasting dolphins for large tankers have made use of panels of German heavy duty Piene piles.

Maplin, 1963–1980

THE THAMES ESTUARY DEVELOPMENT COMPANY

In tracing the Maplin saga it can be seen how a miner's strike, and the three-day week which followed, leading to a change in Government, halted the one big chance that the Port of London had of rivalling Rotterdam as the European major port.

As far back as 1963 the PLA had followed Rotterdam in thinking that 'it is no longer useful to plan for what one thinks will be required; the plan must be for the greatest which can be achieved'. A 90 ft depth of water could be provided at a cost which the traffic could bear. The area considered for development was that reserved in 1964 when Parliament extended the seaward limits of the Port to take in the Maplin Sands adjoining Foulness Island.

In those days the extension of Tilbury dock had just been decided on, and the container revolution had not yet taken place. However, the importance of recovering a large area for future port operations supporting and supported by an industrial complex was foreseen.

Part of the area had been considered as one of many for London's third airport, but this was not a deciding factor in the selection of the area for the development of a deep water port. From the national and port point of view the essentials for development were:

(a) deep water access up to 90 ft at mean High Water Neap Tides;
(b) at least 25 000 to 30 000 acres of land;
(c) port customers who with this specification would guarantee trade; and
(d) landward communications of adequate capacity.

The first two had already been assured with limited customer support, and the land could be created without the airport, especially if an industrial complex shared part of the cost. If it was decided to include an airport, then the costs could be shared between the three ventures.

Early in 1969, Lord Simon, the Chairman of the PLA, announced the formation of the Thames Estuary Development Company Limited (TEDCO). The Company

124

was unique in that it comprised a statutory port authority, the PLA, a local authority, the County Borough of Southend, and private enterprise, represented by two firms of civil engineers – John Howard & Company Limited and John Mowlem & Company Limited – and by London and Thames Haven Oil Wharves Limited, later taken over by Shell UK. Two leading banking houses, Lazard Brothers & Company Limited and Hambros Bank Limited, also agreed to co-operate. This was in keeping with the best traditions of the Thames, as for 400 years companies of merchant adventurers comprising city financiers, corporations, ship owners and merchants had undertaken spectacular new ventures.

The object of the exercise was to plan for the development of the Maplin Sands, to examine in depth the practicability of dredging a deep water approach to the outermost limits of the Port of London, and to reclaim large areas of land upon which the new port facilities could be constructed.

Such a major port development required the consent of the Minister of Transport under Section 9 of the Harbours Act 1964. Industrial development would be a matter for Government in the light of regional planning policy. It was decided to apply for such consent, as it was felt that container ships of the third generation would be so large that a deep water dock would be essential to receive them. An approximate estimate of the cost of the project, carried out at the time, showed that the reclamation of between 7000 and 8000 acres would cost about £50 million. This would allow for an airport with four runways of 12 500 feet, considerably longer than those at Heathrow. It was thought that raising the general level to + 12 OD and surrounding the area with flood walls 30 ft high would cost about £5000 per acre. On the other hand, the land would have been recovered from the sea without the loss of any valuable agricultural land, a tremendous benefit. It would also be necessary to move the Ministry of Defence experimental firing activities, but the cost of doing this, it was thought, would be more than covered by the sale value of the land which this would realize. The financing of the project as a private enterprise would be from the city, not the Government, so the taxpayer would not be expected to meet the bill.

PROPOSED OIL TERMINAL

In 1972 the dredged channel up to the Mobil and Shell refineries at Coryton and Shellhaven was 48 ft deep, which meant that Very Large Crude Carriers (VLCCs) could only come in partly loaded (about 70 000 tons had to be unloaded at sea).

It was proposed therefore that one of the first developments would be the construction of an oil terminal capable initially of taking 250 000 ton tankers fully laden. This would mean a stage 1 dredging of the channel to a depth of 65 ft. Later stage 2 dredging would give 72 ft draught and finally stage 3 would give 85 ft draught, capable of taking 500 000 ton tankers.

A breakthrough in the pumping of crude oil without preheating had been achieved so that it was considered important to lay a pipeline from the terminal to the refineries at Coryton and Shellhaven. It would cross the river to the Isle of Grain

125

BP refinery and finally reach a proposed new refinery at Cliffe. One jetty with a conventional T-head would be built initially, being followed by two or three jetties with the growth of traffic. A buffer storage tank farm would have a capacity of 650 000 cubic metres of crude oil. A throughput of up to 30 million tons in the first year was considered, with twice as much in ten years time. It was hoped that dredging would commence in 1973 and that the oil terminal would be operational in 1976.

The existence of huge deposits of high grade iron ore in Australia, Africa and South America, all very distant from Northern Europe, meant that very large bulk ore carriers were needed to transport this commodity half-way around the world at acceptable costs. Although there are no large steelworks in Southeast England it was considered that there was a case for constructing one to meet the steel strip requirements of the motor industry. Large amounts of scrap generated in the London area could also be used. A mill sited at Maplin would also have had easy access to the North European industrial markets, and an integrated iron and steel works of 10 million tons capacity could easily be accommodated.

MODEL TESTS

One of the early moves was to build a large scale model of the outer estuary to measure the physical effect of such a massive dredging and reclamation project, as it was also necessary to find out whether a channel 90 ft deep would remain open without unacceptable amounts of maintenance dredging.

A large shed was available in Surrey Dock and it was decided to use it for building two models. The first with a horizontal scale of 1:1000 and a vertical scale of 1:100 was used to demonstrate that the shape of the reclamation was hydraulically sound and did not produce unacceptable effects on the regime of the estuary. The second model represented a much larger area, 2000 square miles in all. It covered the coastline from Orford Ness in Suffolk to Margate in Kent with the eastern limit some 50 miles from Southend. Each of the rivers flowing into the estuary was modelled to its tidal limits. The model held up to 4000 cubic feet of water. It was constructed by the Hydraulics Division of the Central Laboratory of George Wimpey & Company Limited, and operated by them. Scientists from the Hydraulics Research Laboratory at Wallingford, who had done such good work previously on the silt problems in the Thames, were responsible for analysing the results. From the tests they were able to predict the long-term effects of reclamation on the estuary for any given shape of land. The hydraulic studies and investigations were vigorously pursued as were the dock layout, wave protection and navigational aspects.

Owing to the intense opposition to alternative sites for the third London Airport, the Government became actively interested in the Maplin scheme. They set up the Maplin Development Authority to study the project and to continue with the trials. TEDCO was no longer needed, and so having served its purpose was disbanded.

Nevertheless, the PLA co-operated closely with Government Departments on all aspects of the project and early in 1973 became Agents for the construction of a trial bund, the next stage in the investigations.

126

PLA

Maplin trial embankment

SITE TESTS

Planning was getting well ahead, but an important factor had to be settled. Would the gravel stay in its dumped position without protection or would serious erosion take place? The only satisfactory answer to the question was to build a trial bank and monitor any movements. By this time the Third London Airport Directorate of the Department of the Environment had been set up and, acting as Agents, the PLA let a contract to the Fairway Consortium. This was formed by Sir Alfred McAlpine Zanen Dredging Company Limited, DOS Dredging Company Limited and Dredging VO2.

The work involved the building of a bank 280 m long, 6 m high and with a top width of approximately 30 m, 2 m above high water level. A trailer suction dredger was used to dredge over 200 000 m³ of gravel from the Thames estuary and a further 13 000 m³ of larger stone taken off-shore near Hastings. When sufficient material had been dumped to appear above high tide level, bulldozers and a large NCK dragline were put ashore to haymake the material to the required height, and to shape it to the correct slope.

SURVEY TOWER

The Property Services Agency designed a survey tower and observation platform to be constructed about 5 km off the Essex coast and 4½ km south of the mouth of Havengore Creek. This was to be used by staff from the Hydraulics Research Station, Wallingford to observe tidal levels and currents, and to sample the silt content of the water, and the platform on the top of the tower would be used as a fixed survey station for setting out and levelling when the Maplin construction got under way.

The steel structure standing 10.4 m above high water spring tides had a top platform 8 m by 5 m for surveying equipment. Below on a second platform some 7.7 m above high water were the living quarters for the survey staff. These were equipped with bunks, cooking and washing facilities. A separate generator building supplied the current. There was also a store room, instrument room and toilet.

The £67 000 contract required a tight programme of thirteen weeks. It was decided that this completion date could only be met by carrying out as much prefabrication work as possible on a quayside site in West India Dock. The piles were welded-up, grit blasted and painted at the same time as the upper section of the tower was fabricated. The hutting was made in the depot and assembled and equipped in West India Dock.

The pontoon *Devon* equipped with its 20-ton derrick was commissioned to carry out the transport work. Not only did it have to take the upper portion of the tower, fender units and four piles each 30.5 m long, but also all the welfare facilities for the men employed on the operation, leaving no spare room at all. On 16 June 1973 the pontoon left West India Dock and was towed to the site. By 20 June the position of the survey tower had been set and checked by the PLA Hydrographic Department, and approved by the Department of the Environment and Hydraulic Research Station Officers.

As the nearest borehole to the site was some distance away there was some uncertainty as to the nature of the sea bed and the underlying strata. It is difficult and time consuming on this type of project to extend the piles in situ, and it was therefore necessary to predict the expected toe level. Before going to sea it was decided that 30.5 m piles would be required, leaving a margin of some 2.5–3 m for variations in strata. This assessment proved to be remarkably accurate as the cut-off was 2.5 m and no delay to the work for extending piles was necessary.

The tower had to be positioned accurately on the top of the piles and it was necessary to devise a means of support during driving. This was achieved by the ingenious method of using the upper portion of the tower as a jig supported on military trestling fitted with adjustable camel's feet sitting on the sea bed. These allowed the structure to be accurately levelled from the irregular sea bed. When this had been done, using parts of the permanent steelwork as guides it was possible to pitch and drive the piles to the required rake of 1 in 6.5. When surveyed the tops of the piles were found to be within 2.5 cm of their correct position. It was then possible, using the derrick, to lift the tower and lower it onto the tops of the piles. Final connections

128

Maplin survey tower leaving West India Dock

Mowlem

were made by bolting and welding, the whole venture taking in all seven days.

Now, six years later, reports show that the trial bank is standing up well with no undue erosion and the tower is being used from time to time for monitoring tidal currents.

MAPLIN TO GO AHEAD

The Maplin Development Act was passed through Parliament and received the Royal Assent in October 1973. Everything was set for the project to go ahead. The oil terminal could be operational by 1976 and the planned opening date for the unit load complex had been set at early 1978, depending on the outcome of the Government's review. At the same time it was hoped that adequate road and rail approaches to the seaport, crucial to the operational date, would be provided. Uncertainties as to the routes to be taken had to be resolved quickly.

Eldon Griffiths, MP, Joint Parliamentary Under Secretary of State for the Environment, in a speech in the autumn of 1973, summed up the position as he saw it at the time 'All big capital projects need to be subjected to critical analysis and prudent judgement. . . . The decision in favour of Maplin rests firmly on the grounds of national need, social responsibility and regard for the future environment. . . . It would also produce a seaport located on the edge of deep water so that fewer but larger ships will be able to use it without having to negotiate narrow waters. This will add to the safety of shipping'.

During the winter of 1973, and the spring of 1974 the information required by the Maplin Development Act was being assembled. However, the miners' strike leading to the downfall of the Government caused some delay. When the new Government under Mr Harold Wilson came to power, a reappraisal of the airport project was initiated. No agreement as to its continuance was immediately given. In the meantime jumbo jets had been introduced on the world's major airlines with considerably larger seating capacity than previously. The Government, seeing this as a means of pacifying environmentalists, thought it best to cancel Maplin Airport. They believed that Heathrow and Gatwick would be able to satisfy the country's needs for the foreseeable future using the new larger planes.

The PLA considered that the Maplin seaport could stand on its own feet. It could become part of a combined scheme with the airport reclamation carried out at the same time or later.

FAILURE TO PROCEED

The discovery of large quantities of North Sea oil almost changed the thinking about an oil terminal overnight. VLCCs would no longer be needed to import large quantities of oil. Certainly the 500 000 ton supertanker for which the seaport was ultimately planned would never be required. It was hoped that by the 1980s the UK would be self-supporting in oil or possibly even an oil-exporting nation. Oil would come ashore by pipeline, or perhaps the Thames-side refineries would be supplied by medium sized tankers bringing the oil around the coast from Scotland.

With the need for a new deep water oil terminal no longer urgent, and a Government unwilling to commit public money at a time of national financial crisis, the PLA had no alternative but to shelve the scheme. Instead they turned to the development of Northfleet Hope for larger container liners.

It is expected that one day, perhaps in the 1980s, the go-ahead will be given for the Maplin port. The PLA have spent nearly £1 million on planning and investigations and it is hoped that all this work is not wasted but merely postponed. The pity is that any future development will cost considerably more than the original scheme.

Dockland development, 1970–1980s, and conclusions

CLOSURE OF ST KATHARINE, LONDON AND SURREY DOCKS

As soon as the new Tilbury Dock began handling a large number of containers, almost every other port decided that it must have a container crane and special berths. The rapid growth of the container trade took everyone by surprise. Inevitably London lost some of its trade to smaller ports around the coast. Southampton followed Tilbury by building a large container complex. Shoreham developed the whole of its Harbour Estate, specializing in timber and bulk wine berths. Folkesone built a roll-on/roll-off facility. Dover carried out major expansions. Sheerness and the Medway ports were developed for civilian use. Felixstowe developed a large container complex. Even Ipswich and Harwich carried out extensions.

As a result of this coastal development it became unnecessary and unfashionable for the short sea trade to come right up the Thames to London when adequate ports had been provided on the coast. The PLA acknowledged this, and decided to close down St Katharine Docks in August 1968, and London Dock in 1969.

The timber trade had turned to bulk carriers with packaged timber and Surrey Dock had found it impossible to handle such large ships. It was decided therefore that this dock should be closed in 1970. Suddenly a total of 350 acres of derelict dockland became available for redevelopment. Regrettably a 160 year old connection with the past was lost.

Much civil engineering history had been built into the docks and would not be seen again. The memories of seeing men bottling port by candle-light in the old London Docks would soon be lost. The coopers checking and repairing wine barrels had already disappeared as bulk wine vats were installed, and the contents transported by road tankers to bottling factories.

ST KATHARINE DOCK REDEVELOPMENT

The site of St Katharine dock, some 25 acres in extent, was sold to the GLC for £1.7 million and a competition was held for a scheme to redevelop the whole area. This was won by Taylor Woodrow, who submitted plans for an ambitious £30 million development in keeping with the history of the dock and its proximity to the Tower of London.

The scheme envisaged an integrated world trading community combining social life and commerce to revitalize this part of the Thames. As there were just over 10 acres of water and a river frontage of 850 feet, there would be ample docking facilities to give London its first yacht basin. The entrance lock, rebuilt in 1958 and equipped with modern 'flap' gates was well suited to serve the dock. There are now berths for 240 craft, either power or sailing, and the moorings can have water, electricity and telephones supplied.

Overlooking the haven is the large Ivory House built in 1854 as the centre of the London ivory trade. This escaped serious damage in World War II, and has been restored at a cost of £1.5 million. The upper floors have been turned into prestigious service apartments and the quay level has been enclosed for shops and a yacht club.

On the southwest corner of the site, once occupied by tea warehouses, the 826-room Tower Hotel has been built. It is a bulky structure, and its size rather overwhelms the dock. However, it gives its guests a wonderful view of the Tower of London, the City, the Thames and the dock itself.

The original development scheme was intended to take nine years to complete, but a major public reappraisal of the whole project, particularly the amount of housing to be incorporated, delayed completion by several years.

One of the more unusual building is the *Dickens Inn*, a four-storey timber-framed building discovered inside two outer layers of brickwork during the demolition of an old warehouse. Industrial archaeologists believe it to have been used as a

Bulk wine vats, West India Dock

brewery at least as early as 1800, and it was thought that it must have been in too good a condition to demolish during the building of the dock. The complete wooden structure was jacked up and rolled 75 yards to its new position beside the eastern basin. Extensive renovation has produced a pleasing building, housing two restaurants and a bar.

Taylor Woodrow and the Marine Trust have created a floating exhibition housed in the eastern basin. Historic vessels, some having long associations with the Thames, have been collected together, among them a three-masted schooner built in 1900, one of the last to trade with the West Country, and the *Nore*, one of the light-ships which used to be moored in the estuary. A number of Thames sailing barges have also been berthed in the basin.

The southeast side of the dock was allocated for housing and 300 flats have been built for the GLC. Besides the council flats there are plans to build 400 private homes.

The west side of the dock was occupied by the Mint Bonded tea warehouses which ran for 500 ft alongside Tower Bridge approach. The warehouses were damaged by bombing in 1940, repaired in 1949, and since the closure of St Katharine dock had remained empty, awaiting redevelopment. Decay had set in, however, and this, coupled with recent arson, has made repair impossible so they have had to be demolished. Rebuilding is to designs resembling Hardwick's orig-inal Victorian building. The new building is to be named International House and on completion will form the second stage of the World Trade Centre.

The first stage of the London World Trade Centre is housed on the northwest corner of the dock. It occupies part of the site of the old St Katharine Dock House, completely destroyed by enemy action in World War II, and partly the western half of C warehouse demolished in 1959. St Katharine Dock House was built by John Mowlem & Company Limited in the early 1960s to house the Accounting and Com-puter centre for the PLA. Designed by Andrew Renton and Partners, it won the Architectural Award for 1967. On the sale of the Trinity Square Head Office in 1970 the PLA management leased the top floor after selling the whole building to Taylor Woodrow for the World Trade Centre.

The World Trade Centre Association links 45 countries and has more than 1000 members, some of whom have permanent offices there. Others can use the facili-ties for communications, research offices and club services.

The International House is to be linked to the Tower Hotel and World Trade Centre and to Tower Hill Underground Station by means of a subway being incor-porated in the new road layout at Tower Hill.

The presence of the World Trade Centre and the Tower Hotel in the St Katharine Dock development has led to an eastwards expansion of the City's activities. New offices are being built near St Katharine's Dock House in an area until recently classified as slum.

The final details of the redevelopment of the northeast corner have not yet been released. The last traces of the old dock will soon vanish when the remaining section of C warehouse is demolished. One of the suggested uses of the area is the establishment of a World Commodities Centre, with an adjoining sports complex.

St Katharine Dock, being only 25 acres in extent, was of a suitable size to lend itself to a single development bearing in mind that 10 acres is covered by water. The success of the enterprise can be measured by the fact that over two million people a year are visiting the Maritime Museum and admiring the tasteful manner in which Taylor Woodrow have retained an old world atmosphere in a corner of a modern Trade Centre.

DOCKLANDS DEVELOPMENT PLAN

In April 1971 the Department of the Environment with the Greater London Council commissioned the Consulting Engineers R. Travers Morgan to carry out a comprehensive study of the docklands area. The study area, some 8½ square miles in extent, housed about 55 000 people and provided about 60 000 jobs in 1971. The area included all the upstream docks. The Consultants put forward a number of proposals such as:

(a) the provision of continuous small scale water areas based on the existing docks, and the creation round them of an urban environment with extensive areas of housing;

(b) the creation of major parklands on the scale of Greenwich Park, but partly wooded and with substantial water areas;

Ivory House, St Katharine Dock

134

Tower Hotel, St Katharine Dock

(c) a transformation of the area with an emphasis on owner-occupied housing, including more expensive houses, together with major office centres, but little industry;

(d) reinforcement of the traditional East London role, with provision of new industrial employment and extensive housing for lower income groups, but with greatly improved residential environment without high rise flats; or

(e) a fairly intensive development with equal quantities of rented and owner occupied houses, new office employment, extension of existing industry and the creation of a major shopping centre.

The cost of the schemes varied between £450 million and £800 million exclusive of the cost of land. It was assumed that most of the redevelopment would take place between 1978 and 1991.

The report, submitted in January 1973, assumed that the PLA would have moved downstream to Maplin and Tilbury by 1978. This would leave the Beckton land, the Royals, East India, West India and Millwall, London and Surrey Docks free to be incorporated in their scheme. They thought that West India and Millwall Docks would be closed in 1977, Victoria in 1982, and King George V with Albert Dock in 1987.

The assumed dock closure dates were those of the Consultants: the PLA at that time had made no decisions as to the closure dates. These would be dependent on the continuing requirements for any of the upper docks to remain open, as dictated by the level and pattern of trade. In any case there was still the commitment to keep the flour mills on the south side of Victoria Dock supplied with grain by water.

The 1973 report envisaged schemes too vast and too expensive to be undertaken at one time, so no action was taken for several years.

Trade in the India and Millwall Docks increased and a new Combi berth was opened on the site of the old K warehouse, designed to deal with ships of mixed container and packaged goods, though judging by the amount of trade available, it would have to be doubled in size to break even financially. Trade also would be dependent on the shipping companies' continued willingness to make the costly haul up the River.

ROAD AND UNDERGROUND COMMUNICATIONS

The docklands area is poorly served by both road and rail services, in fact communication is far worse than it was at the turn of the century when many dockside railway stations provided services to the City.

The revitalization of the whole area depends on the extension of the London Transport's Jubilee line eastwards from Charing Cross. The scheme favoured by the GLC in April 1979 was to start at the existing Charing Cross terminal following the north bank of the river to Fenchurch Street. Here the line would branch, the southern branch crossing the river to New Cross and Lewisham, and the eastern branch crossing the river five times with stations at Wapping, Surrey Dock, Isle of Dogs, Greenwich, Canning Town, Woolwich and Thamesmead. The line of the route, at the time of writing, has not yet received Parliamentary approval. Should the scheme go ahead it would go a long way towards making the inhabitants of the Isle of Dogs feel less cut off from the rest of London. It might also soften the feeling of isolation and frustration leading them at times to threaten a state of independence.

DOCKLANDS ROAD PLAN

Schemes for improving roads in the area have been suggested and turned down regularly. In fact no new roads have been constructed since the ambitious start was made in 1934 with the construction of Silvertown Way, joining Silvertown, Victoria Dock and King George V Dock to the A13 at Canning Town.

Congestion along the East India Dock and Commercial Roads has never been relieved. A by-pass from the Canning Town Viaduct through East India and the north of West India Docks joining the now widened Ratcliffe Highway would help considerably.

The GLC has proposed a £130 million southern relief road from the New Kent Road through Southwark Park and Surrey Dock crossing the river by tunnel to the Isle of Dogs, recrossing to Greenwich and joining the M2 motorway at Charlton. The scheme has so far been turned down by the Docklands Joint Committee, and the Southwark and Tower Hamlets Councils.

The Southwark Council were quoted as saying that the road would strangle developments in Surrey Docks where they had plans for housing and industrial development.

SURREY DOCK

About 270 acres in Surrey Dock became available when it closed in 1970. Of the schemes planned, four of the most seriously considered were:

(a) a rail terminal and marshalling yard for the Channel Tunnel;

(b) a trade mart on 135 acres, in which development both the PLA and GLC were to have had an involvement (when the American-based company deferred their involvement in the scheme for an indefinite period the PLA sold the site to the GLC);

(c) a venue for the 1988 Olympics; and

(d) an airport for vertical take-off aircraft.

Unfortunately the world trade recession and the depressed British economy have created an unfavourable climate for the investment needed to get these schemes under way.

The Surrey Commercial Docks area was finally sold to Southwark Borough Council in the summer of 1977 for a housing estate. The Council has been a little slow in making out a programme, but in January 1979 a start had been made on the first 281 houses. It is proposed to provide open spaces and facilities for leisure. In the next few years the Council proposes to spend £¾ million on the creation of a central park with water features, streams, and 10 000 trees. A similar sum will be provided for football and cricket pitches, and an elaborate pavilion. In the Greenland and South Docks there are proposals for the creation of a water-based canoeing and boating area with 150 moorings with private housing nearby.

THE WAPPING PLAN

In January 1977 the Consulting Engineers Ove Arup & Partners produced what was known as the 'Wapping Plan' for the treatment of the old London Dock area.

The main suggestions were to reduce the size and depth of the water area by filling in and lagooning the silt into smaller areas where necessary. The Tobacco dock and Hermitage basin would also be partially filled in and surrounded by public walks and trees. The Eastern Dock has already been filled in, as has Wapping Basin. They cannot be built over owing to the foundation costs, but will become playing fields planted with trees. The Shadwell basin has been designated as an aquatic recreational area, but it will be necessary to introduce some degree of water circulating together with impervious dams at the entrances and a beach. The depth of water would be reduced to about six feet. Most of the old vaults will be demolished, but some of the old boundary walls may remain as a reminder of the old dock system.

A feature of the area will be the Riverside Walk with access to the river provided wherever possible. The first phase of the housing development has already started with John Laing Limited building two- or three-storey houses and low groups of small flats on the north side of the Eastern Basin.

The scheme when completed should form an attractive area, with many open spaces. Although the politics of London boroughs have resulted in a vast preponderance of public sector housing in dockland, the planners are hoping that when schools, shopping centres, roads, and wide open spaces are in evidence, private investment will follow.

BECKTON MARSHES

The land to the north of Albert Dock, which Sir Frederick Palmer wanted to develop in 1910, and which for nearly 70 years has remained an open space, was finally handed over to the London Borough of Newham for their housing pro-

Surrey Canal, 1975

gramme. An elaborate system of land drainage has been installed, with a surface water pumping station situated near to the old Gallions Hotel, and is now in operation. A foul pumping station is shortly to be constructed near the Beckton bypass. When road and drains have been installed, a housing scheme can be started, thus freeing some of the older parts of the Borough for redevelopment.

The Government was not satisfied with the progress made by the various local authorities and has decided to set up an Urban Development Corporation. This will be responsible for producing strategic plans, and for drawing together planning and control functions of existing bodies, and the Docklands Joint Committee. The new Corporation will have wide powers, and will be controlled by 'dynamic' people. The formation of this body has been welcomed by the GLC, which is keen to see the redevelopment ball begin to roll.

CONCLUSIONS

This book has been written as a tribute to all the Engineers who for 180 years planned, developed, and constructed new works to keep London the premier port in the world. However, in recent years changes have taken place so rapidly that soon the older docks will have vanished almost without trace. The old London Docks were too far upstream for the new container and bulk cargo carriers which required both a quick turnround, and deep water.

The building of the Thames Barrier in Woolwich Reach has led to an increase in insurance rates for ships passing through. This has been one of the factors influencing the decision of the Port of London Authority, taken in March 1980, to close both the West India and Millwall Docks. Earlier in the year the PLA had made it quite clear that these docks could be saved only with determined efforts which included ending restrictive practices and full co-operation in the fight to cut costs and win new business. A strike in support of a pay claim accelerated the closure and it was decided to transfer the remaining trade to the Royal Docks which are under threat of closure themselves. Tilbury remains the leading container port in the United Kingdom.

When towards the end of 1979 the question of the third London Airport was again raised, there was considerable interest in the revival of the Maplin air/seaport project. Both the Greater London Council and the Essex County Council favoured Maplin, and appointed Sir William Halcrow and Partners to examine the advantages and disadvantages of the scheme. However, the Government made it quite clear that money would not be made available for such a project at a time of stringent economy. The Maplin Port upon which the PLA had pinned its hopes for the restoration of London's greatness will once again be delayed.

Bibliography

Papers on London Docks published in the *Proceedings* of the Institution of Civil Engineers

Year	Author	Title	Volume and page
1859	W. J. Kingsbury	Description of the Entrance, Entrance Lock, and Jetty Walls of the Victoria (London) Docks	**18**, 445
1871	L. F. Vernon-Harcourt	Description of the new South Dock in the Isle of Dogs, forming part of the West India Docks	**34**, 157
1895	J. F. Scott	The construction and equipment of the Tilbury Docks	**120**, 276
1909	M. Mowat	Some recent grain-handling and storing appliances at the Millwall Docks	**177**, 58
1922	F. M. G. Du-Plat-Taylor	Extensions at Tilbury Docks, 1912–1917	**215**, 165
1923	A. Binns	The King George V Dock, London	**216**, 372
1931	Sir Cyril Kirkpatrick	Presidential Address: The Tidal Thames	**233**, 1
1935	F. W. D. Davis and W. MacKenzie	Major improvement-works of the Port of London Authority, 1925–1930	**240**, 258
1939	R. R. Liddell	Improvements at the Royal Docks, Port of London Authority	**10**, 283
1953	G. A. Wilson	Port of London Authority Engineering Works, 1952	Part II, **2**, 551
1954	W. P. Shepherd-Barron	Presidential Address: The docks of London	Part I, **3**, 12
1954	N. N. B. Ordman and I. S. S. Greeves	Design and construction of a pre-stressed concrete framed transit shed for the Port of London Authority	Part III, **3**, 409
1956	J. A. Fisher	Reconstruction of the Gallions Lower Entrance Lock at the Royal Docks of the Port of London Authority	Part II, **5**, 136
1957	Sir Charles Inglis	The Regimen of the Thames Estuary as affected by currents, salinities, and river flow	**7**, 827

Year	Author	Title	Volume and page
1957	C. Peel, A. J. Carmichael and R. F. J. Smeardon	No. 1 Berth, Tilbury Dock	**8,** 331
1960	D. E. Glover, E. Newton, H. M. Dale and T. R. Brown	Port of London Authority: development of two dock areas, 1959	**15,** 411
1961	R. Carey and C. G. Cumming	The design and construction of Erith Jetty	**18,** 15
1967	N. N. B. Ordman	Port planning: some basic considerations	**37,** 257
1967	R. F. J. Smeardon, E. Newton and F. A. Page	Engineering works at Tilbury Docks 1963–67	**38,** 177
1974	A. Beckett and M. Pudden	A high capacity container terminal	**56,** 161

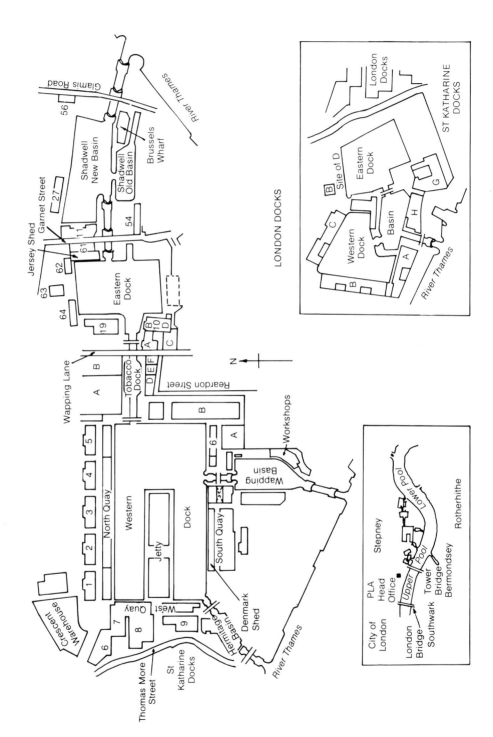

London and St Katharine Docks

142

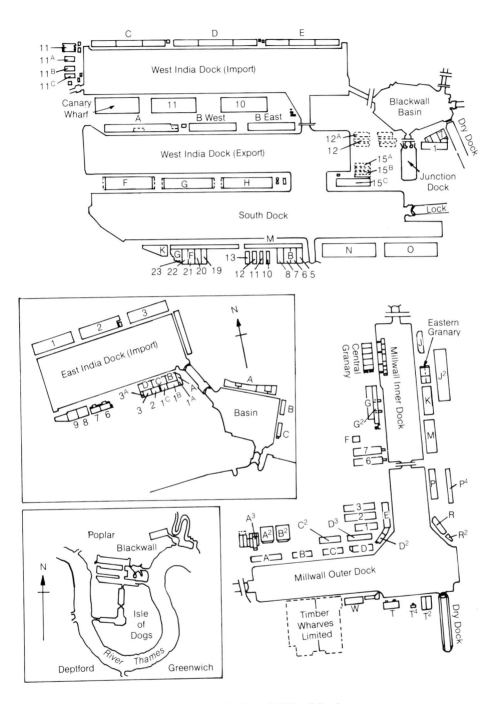

East and West India and Millwall Docks

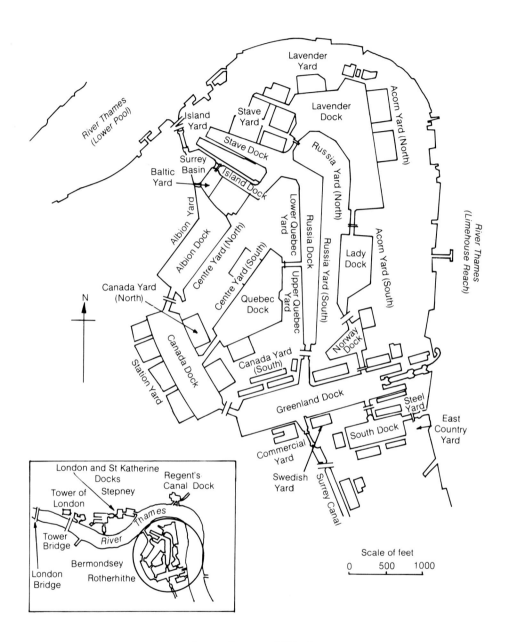

River Thames (Lower Pool)

Lavender Yard

Lavender Dock

Acorn Yard (North)

Island Yard

Stave Yard

Stave Dock

Surrey Basin

Baltic Yard

Island Dock

Russia Yard (North)

Albion Yard

Albion Dock

Centre Yard (North)

Centre Yard (South)

Lower Quebec Yard

Russia Dock

Russia Yard (South)

Lady Dock

Acorn Yard (South)

River Thames (Limehouse Reach)

Canada Yard (North)

Upper Quebec Yard

Quebec Dock

Norway Dock

Canada Dock

Station Yard

Canada Yard (South)

Greenland Dock

Steel Yard

East Country Yard

South Dock

Commercial Yard

Swedish Yard

Surrey Canal

N

Scale of feet

0   500   1000

London and St Katherine Docks

Regent's Canal Dock

Tower of London

Stepney

Tower Bridge

River Thames

London Bridge

Bermondsey

Rotherhithe

Surrey Commercial Docks

144

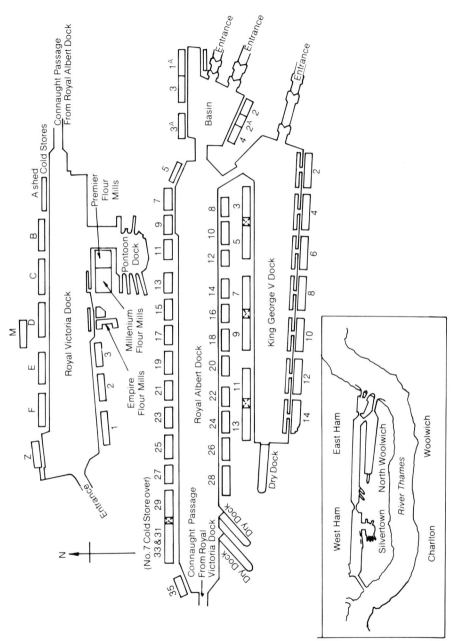

Royal Victoria and Albert and King George V Docks

145

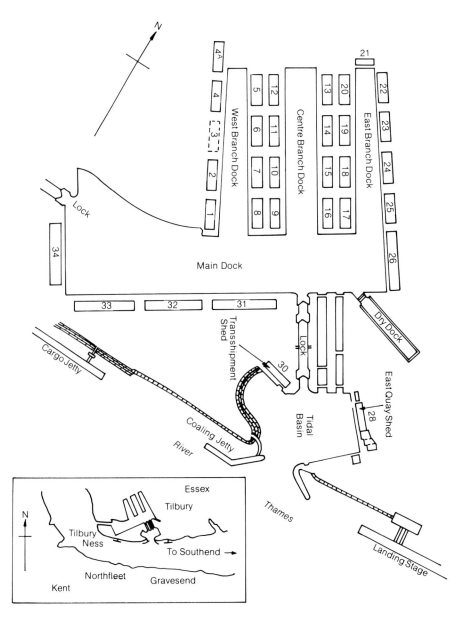

Tilbury Docks (before recent developments)

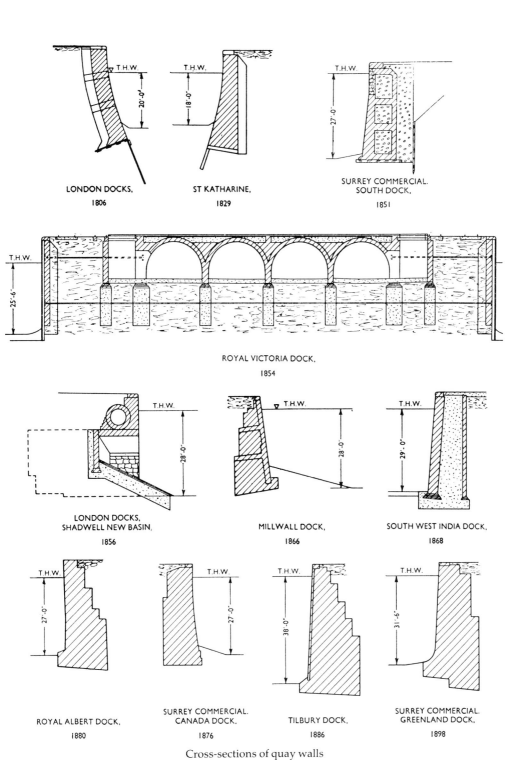

LONDON DOCKS,
1806

ST KATHARINE,
1829

SURREY COMMERCIAL.
SOUTH DOCK,
1851

ROYAL VICTORIA DOCK,
1854

LONDON DOCKS,
SHADWELL NEW BASIN,
1856

MILLWALL DOCK,
1866

SOUTH WEST INDIA DOCK,
1868

ROYAL ALBERT DOCK,
1880

SURREY COMMERCIAL.
CANADA DOCK,
1876

TILBURY DOCK,
1886

SURREY COMMERCIAL.
GREENLAND DOCK,
1898

Cross-sections of quay walls

147

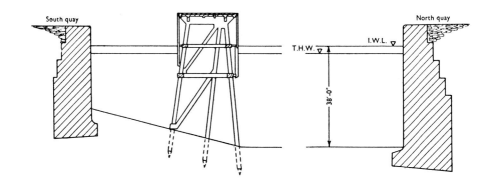

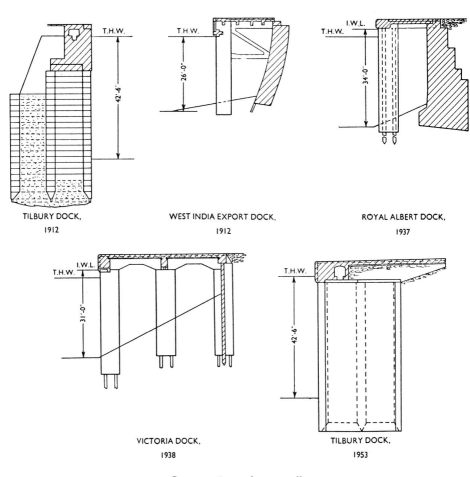

KING GEORGE V DOCK,
1912-1921

TILBURY DOCK,
1912

WEST INDIA EXPORT DOCK,
1912

ROYAL ALBERT DOCK,
1937

VICTORIA DOCK,
1938

TILBURY DOCK,
1953

Cross-sections of quay walls

148

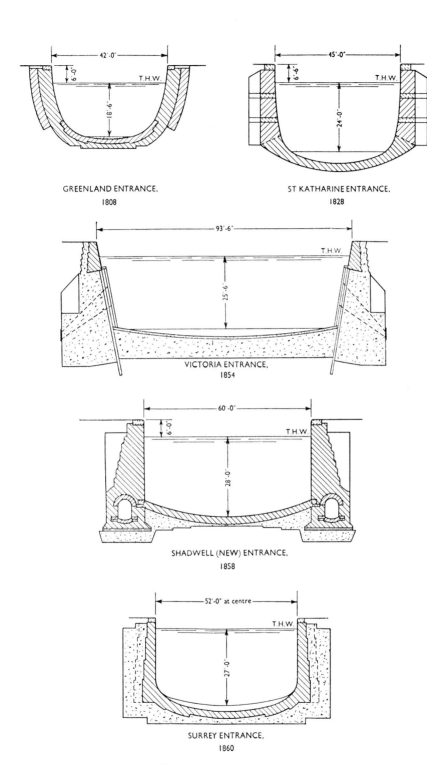

GREENLAND ENTRANCE,
1808

ST KATHARINE ENTRANCE,
1828

VICTORIA ENTRANCE,
1854

SHADWELL (NEW) ENTRANCE,
1858

SURREY ENTRANCE,
1860

Cross-sections of entrances

149

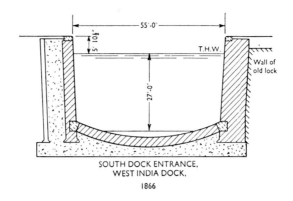

SOUTH DOCK ENTRANCE,
WEST INDIA DOCK,
1866

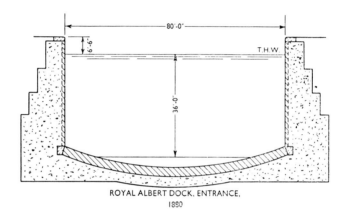

ROYAL ALBERT DOCK, ENTRANCE,
1890

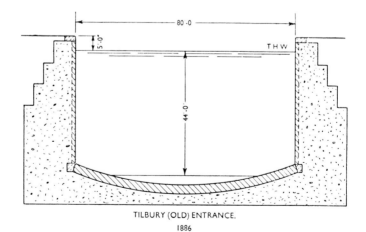

TILBURY (OLD) ENTRANCE,
1886

Cross-sections of entrances

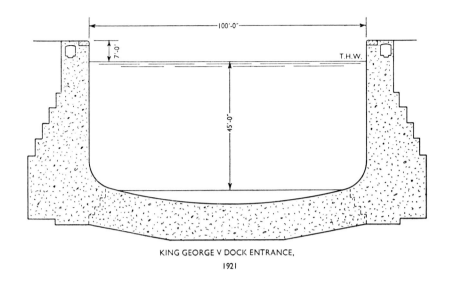

KING GEORGE V DOCK ENTRANCE,
1921

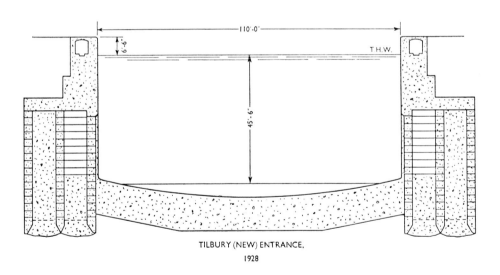

TILBURY (NEW) ENTRANCE,
1928

Cross-sections of entrances

151

Index

154

DATE DUE

MA

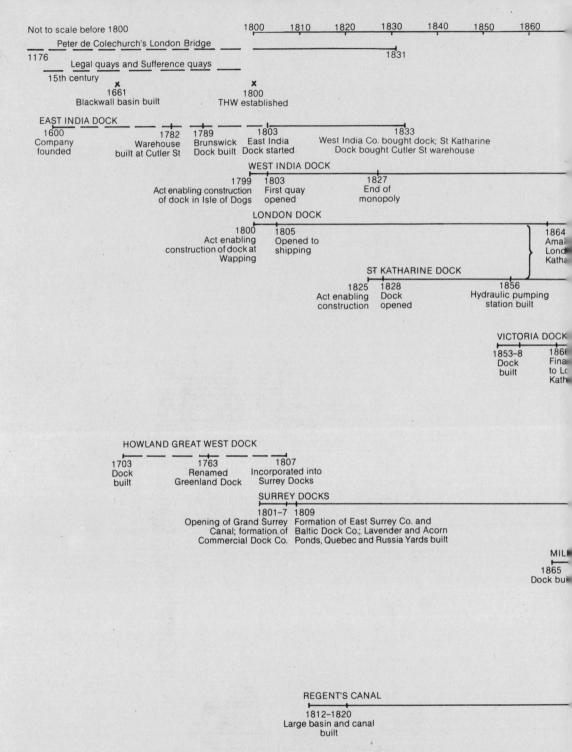

Not to scale before 1800

| | 1800 | 1810 | 1820 | 1830 | 1840 | 1850 | 1860 |

Peter de Colechurch's London Bridge
1176 ... 1831

Legal quays and Sufference quays
15th century

✗ 1661
Blackwall basin built

✗ 1800
THW established

EAST INDIA DOCK
1600
Company
founded

1782
Warehouse
built at Cutler St

1789
Brunswick
Dock built

1803
East India
Dock started

1833
West India Co. bought dock; St Katharine
Dock bought Cutler St warehouse

WEST INDIA DOCK
1799
Act enabling construction
of dock in Isle of Dogs

1803
First quay
opened

1827
End of
monopoly

LONDON DOCK
1800
Act enabling
construction of dock at
Wapping

1805
Opened to
shipping

1864
Ama
Lond
Katha

ST KATHARINE DOCK
1825
Act enabling
construction

1828
Dock
opened

1856
Hydraulic pumping
station built

VICTORIA DOCK
1853–8
Dock
built

1866
Fina
to Lo
Kath

HOWLAND GREAT WEST DOCK
1703
Dock
built

1763
Renamed
Greenland Dock

1807
Incorporated into
Surrey Docks

SURREY DOCKS
1801–7
Opening of Grand Surrey
Canal; formation of
Commercial Dock Co.

1809
Formation of East Surrey Co. and
Baltic Dock Co.; Lavender and Acorn
Ponds, Quebec and Russia Yards built

MILI
1865
Dock bui

REGENT'S CANAL
1812–1820
Large basin and canal
built

SCHEMATIC HISTORY OF LONDON DOCKS